# HOW AC & DC CIRCUITS WORK

BASIC

ELECTRICITY/ELECTRONICS

## VOLUME 2

# HOW AC & DC CIRCUITS WORK

By Training & Retraining, Inc.

TRI

Howard W. Sams & Co., Inc.

4300 WEST 62ND ST. INDIANAPOLIS, INDIANA 46268 USA

FIRST EDITION
THIRTEENTH PRINTING—1979



International Standard Book Number: 0-672-20168-2
Library of Congress Catalog Card Number: 63-23002

*Printed in the United States of America.*

# Acknowledgments

Grateful acknowledgment is made to all those who participated in the preparation, compilation, and editing of this series. Without their valuable contributions this series would not have been possible.

In this regard, prime consideration is due Bernard C. Monnes, Educational Specialist, Navy Electronics School, for his excellent contributions in the areas of writing, editorial organization, and final review of the entire series. The finalization of these volumes, both as to technical content and educational value, is due principally to his tireless and conscientious efforts.

Grateful appreciation is also extended to Lt. Loren Worley, USN, and Ashley G. Skidmore, BUSHIPS, Dept. of the Navy, for their original preparatory contributions and coediting of this series. We also want to thank Irene and Don Koosis, Raymond Mungiu, George V. Novotny, and Robert J. Brite for their technical writing and contribution to the programmed method of presentation. Special thanks to Robert L. Snyder for his initial preparation and organizational work on the complete series.

Finally, special thanks are due the Publisher's editorial staff for invaluable assistance beyond the normal publisher-author relationship.

STANLEY B. SCHIFFMAN, *President,*
*Training & Retraining, Inc.*

# Introduction

This second volume in the series shows how the basic principles of electricity apply to the behavior of AC and DC circuits, the building blocks from which all electrical and electronic equipment is made. As the principles of circuit operation are presented, they are related to familiar applications to help you understand both the principle itself and how it can be put to practical use. Thus the learning process is made easier, and your interest in the subject is maintained. In this volume you will progress beyond basic generalities and begin to learn specific facts of greater practical importance. After studying this volume, you will be able to apply your knowledge of circuit fundamentals to the analysis of how electrical and electronic devices work.

## WHAT YOU WILL LEARN

This volume explains how the relationship between voltage and current in a circuit depends on the arrangement of components in the circuit. You will learn about these components (resistors, inductors, capacitors, transformers, etc.) and study the basic ways in which they can be connected. Methods of calculating the combined effect of several resistors, inductors, or capacitors connected in a circuit are also discussed. The text explains reactance—how inductors and capacitors produce different results for AC than they do for DC. You will discover how reactance and resistance produce a combined effect called impedance. Special combinations of capacitance and inductance that produce a condition called resonance are presented, and you are shown how this condition can be put to use in tuning radio and television receivers, as well as in many other applications. The meaning of phase and the use of vectors when studying the actions of AC in a circuit are explained. You will become familiar with time constants and be introduced to the fundamentals of pulse circuits. Finally, you will study about transformers, how they work, and how and why they are used.

## WHAT YOU SHOULD KNOW BEFORE YOU START

Before studying this text it is desirable, but not absolutely necessary, that you have a general familiarity with the basic principles of electricity and electronics (such as is provided by Volume 1 of this series). However, the only essential prerequisites for learning about AC and DC circuits from this text are an ability to read and a desire to learn. All terms are carefully defined. Enough math is used to give precise interpretation to important principles, but if you know how to add, subtract, multiply, and divide, the mathematical expressions will give you no trouble.

## WHY THE TEXT FORMAT WAS CHOSEN

During the past few years, new concepts of learning have been developed under the common heading of programmed instruction. Although there are arguments for and against each of the several formats or styles of programmed textbooks, the value of programmed instruction itself has been proved to be sound. Most educators now seem to agree that the style of programming should be developed to fit the needs of teaching the particular subject. To help you progress successfully through this volume, a brief explanation of the programmed format follows.

Each chapter is divided into small bits of information presented in a sequence that has proved best for learning purposes. Some of the information bits are very short—a single sentence in some cases. Others may include several paragraphs. The length of each presentation is determined by the nature of the concept being explained and the knowledge the reader has gained up to that point.

The text is designed around two-page segments. Facing pages include information on one or more concepts, complete with illustrations designed to clarify the word descriptions used. Self-testing questions are included in most of these two-page segments. Many of these questions are in the form of statements requiring that you fill in one or more missing words; other questions are either multiple-choice or simple essay types. Answers are given on the succeeding page, so you will have the opportunity to check the accuracy of your response and verify what you have or have not learned before proceeding. When you find that your answer to a question does not agree with that given, you should restudy the information to determine why your answer was incorrect. As you can see, this method of question-answer programming insures that you will advance through the text as quickly as you are able to absorb what has been presented.

The beginning of each chapter features a preview of its contents, and a review of the important points is contained at the end of the chapter. The preview gives you an idea of the purpose of the chapter—what you can expect to learn. This helps to give practical meaning to the information as it is presented. The review at the completion of the chapter summarizes its content so that you can locate and restudy those areas which have escaped your full comprehension. And, just as important, the review is a definite aid to retention and recall of what you have learned.

## HOW YOU SHOULD STUDY THIS TEXT

Naturally, good study habits are important. You should set aside a specific time each day to study in an area where you can concentrate without being disturbed. Select a time when you are at your mental peak, a period when you feel most alert.

Here are a few pointers you will find helpful in getting the most out of this volume.

1. Read each sentence carefully and deliberately. There are no unnecessary words or phrases; each sentence pre-

sents or supports a thought which is important to your understanding of electricity and electronics.

2. When you are referred to or come to an illustration, stop at the end of the sentence you are reading and study the illustration. Make sure you have a mental picture of its general content. Then continue reading, returning to the illustration each time a detailed examination is required. The drawings were especially planned to reinforce your understanding of the subject.
3. At the bottom of most right-hand pages you will find one or more questions to be answered. Some of these contain "fill-in" blanks. Since more than one word might logically fill a given blank, the number of dashes indicates the number of letters in the desired word. In answering the questions, it is important that you actually do so in writing, either in the book or on a separate sheet of paper. The physical act of writing the answers provides greater retention than merely thinking the answer. Writing will not become a chore since most of the required answers are short.
4. Answer all questions in a section before turning the page to check the accuracy of your responses. Refer to any of the material you have read if you need help. If you don't know the answer even after a quick review of the related text, finish answering any remaining questions. If the answers to any questions you skipped still haven't come to you, turn the page and check the answer section.
5. When you have answered a question incorrectly, return to the appropriate paragraph or page and restudy the material. Knowing the correct answer to a question is less important than understanding **why** it is correct. Each section of new material is based on previously presented information. If there is a weak link in this chain, the later material will be more difficult to understand.
6. In some instances, the text describes certain principles in terms of the results of simple experiments. The information is presented so that you will gain knowledge

whether you perform the experiments or not. However, you will gain a greater understanding of the subject if you do perform the suggested experiments.

7. Carefully study the review, "What You Have Learned," at the end of each chapter. This review will help you gauge your knowledge of the information in the chapter and actually reinforce your knowledge. When you run across statements you don't completely understand, reread the sections relating to these statements, and recheck the questions and answers before going to the next chapter.

This volume has been carefully planned to make the learning process as easy as possible. Naturally, a certain amount of effort on your part is required if you are to obtain the maximum benefit from the book. However, if you follow the pointers just given, your efforts will be well rewarded, and you will find that your study of electricity and electronics will be a pleasant and interesting experience.

# Contents

## CHAPTER 1



## CHAPTER 2



## CHAPTER 3

## CHAPTER 4



## CHAPTER 5



## CHAPTER 6



## CHAPTER 7



## CHAPTER 8

## CHAPTER 9



## CHAPTER 10



## CHAPTER 11

## CHAPTER 12



## CHAPTER 13



## CHAPTER 14

# 1

# Understanding Basic Principles

## *What You Will Learn*

**It is important that you learn to visualize and describe electron flow and be able to define the difference between conductors and insulators. In this chapter you will learn to identify and describe the six methods for developing electricity. You will determine the effect of resistance and voltage on current flow and become familiar with actual devices.**

### WHAT IS ELECTRICITY?

Electricity is voltage and current. Voltage is electrical pressure, and current is the flow of charged particles.

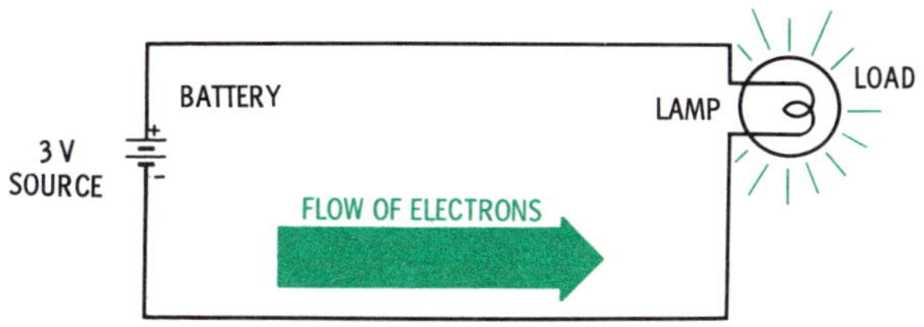


**Voltage and current.**

Voltage (electrical pressure) is an excess of negatively charged particles at one terminal of a source with respect to the other terminal.

Current is the movement of these charged particles from the negative terminal of the source (battery), through the load (lamp), and back to the positive terminal of the source.

## THE MOLECULE

To establish a mental picture of these charged particles and their movement, you must visualize how all matter is put together. First, consider the smallest grain of salt you can see. Assume that we break it in half and then break

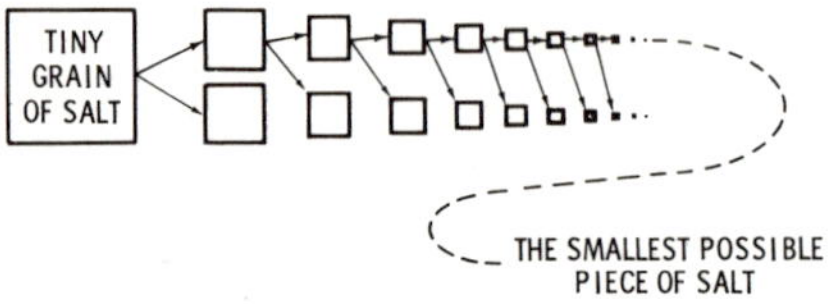


**Division of a grain of salt.**

one of the halves in half, continuing the process until we have the smallest piece of salt possible. This piece cannot be seen, even with the most powerful microscope. This breakdown to the smallest possible piece of salt would require several million of the ½-of-½ steps.

The smallest possible piece of salt is called a **molecule.** It consists of two elements—chlorine and sodium. A molecule, by definition, is the smallest particle of a substance that still retains the same physical and chemical characteristics

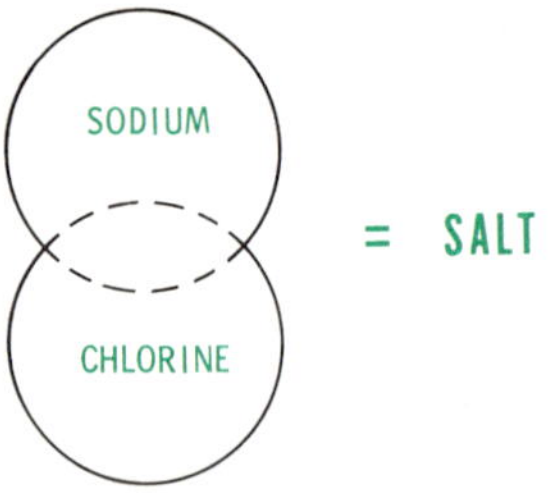


**Salt molecule.**

of the substance. A molecule of salt is made up of two elements—one part chlorine and one part sodium—chemically bonded, or "welded," together. Water is composed of two parts hydrogen and one part oxygen.

## THE ATOM

The elements that are bonded together to form the molecule are called **atoms.** The atom, by definition, is the smallest portion of an element which exhibits all properties of the element. An **element** is one of a class of substances (of which more than 100 are now recognized) which cannot be separated into substances of other kinds. The atom is made up of even smaller particles called **protons, electrons,** and **neutrons.** There are other tiny particles in an atom, but these three are all that will be discussed.

**EQUATIONS FOR SALT AND WATER**

| | | |
|---|---|---|
| ONE MOLECULE OF SALT | = | ONE ATOM OF SODIUM + ONE ATOM OF CHLORINE |
| ONE MOLECULE OF WATER | = | TWO ATOMS OF HYDROGEN + ONE ATOM OF OXYGEN |

The helium atom has two electrons in orbit around a nucleus of two protons and two neutrons. An electron has a negative charge that is equal to but opposite in polarity

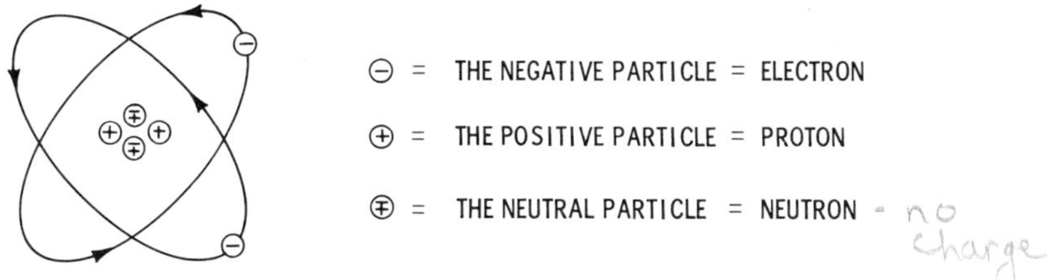


**The helium atom.**

to the positive charge of a proton. The atom is very similar to the solar system in structure.

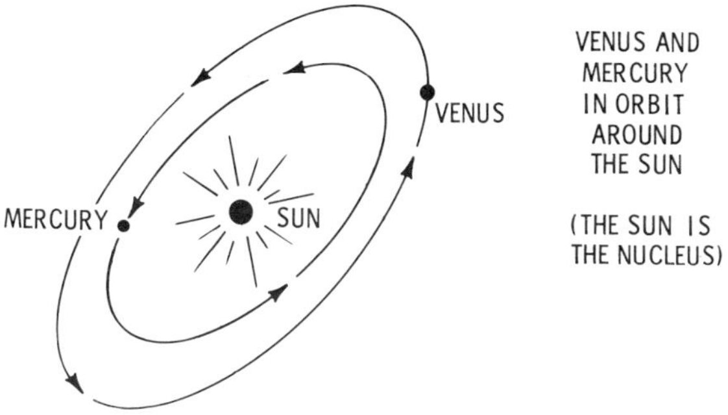


**Part of the solar system.**

**Q1. Of what would the smallest particle of water be made? Draw a diagram similar to the one for a molecule of salt on a separate sheet of paper.**

**Your Answer Should Be:**

**A1.** The smallest particle of water would be made of **two parts hydrogen and one part oxygen.**

**A molecule of water.**

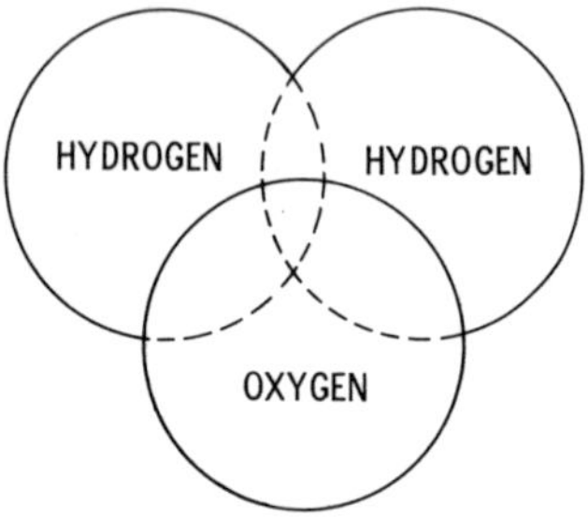


## FREE ELECTRONS

If the proper amount of energy in the form of heat, light, electrical pressure, etc., is concentrated in an atom, it can cause the atom to give up or take on electrical particles. Elements differ from each other by the number of electrons in orbit and how many protons and neutrons are in the nucleus.

Normal atoms have the same number of protons in the nucleus as they have electrons in orbits.

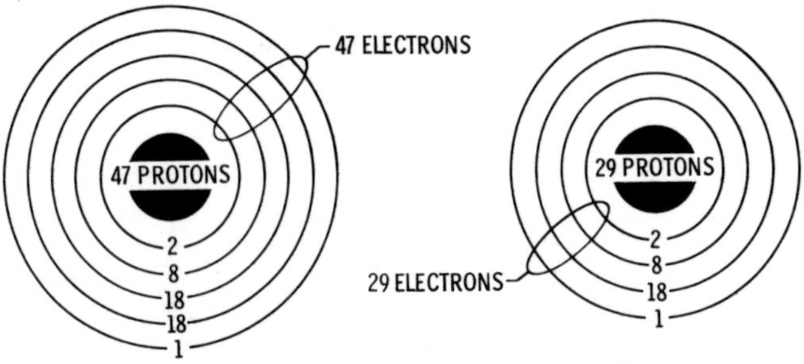


*Silver.* *Copper.*

**Silver and copper atoms.**

Electrons do not all move in the same direction around the nucleus. They travel in many different orbit paths (as many paths as there are electrons). Counting from the nucleus outward, the greatest number of electrons that can exist in the first and second orbit zones are 2 and 8, respectively. Subsequent zones may be filled by 8, 18, or 32 electrons. When the outer orbit zone is not completely filled, the element (atom) has the ability to release **free electrons** when voltage is applied.

Silver and copper atoms have only one electron in their outer orbit zone. The amount of energy needed to move these electrons to a nearby atom will be less than if the outer orbit zone were completely filled.

Some elements have all of their orbit zones completely filled. These elements are called **inert** because they will neither give up nor accept an electron from another atom. The neon atom is a good example of this concept. Its atomic number is 10—it has 10 electrons in orbit around a nucleus containing 10 protons. As can be seen in the illustration below, the outer orbit zone contains 8 electrons, thus it is completely filled. By contrast, the fluorine atom has only 7 electrons in its outer orbit. This makes fluorine an **active** element rather than an inert element.

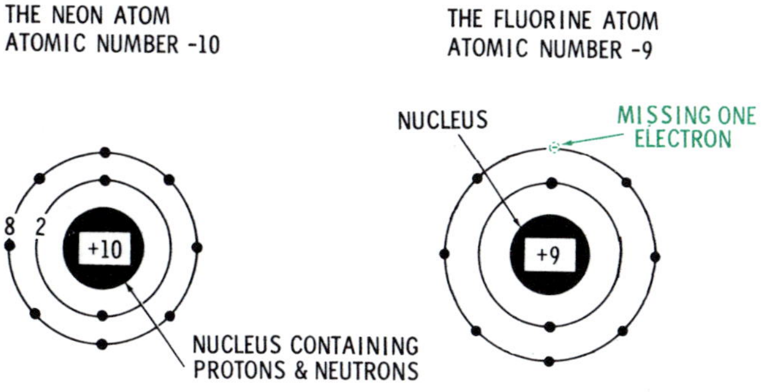


**Inert and active elements.**

**Q2. An atom, like the solar system, is mostly ______.**

**Q3. Could a molecule be a single atom? (Review each definition carefully.)**

**Q4. The electrons go from the negative terminal, through the load, and to the positive terminal of the source. What path do they follow inside a battery?**

**Q5. Name the particles in the nucleus and assign the proper charges. Assign the proper charge to the particles which orbit about the nucleus.**

**Your Answers Should Be:**

**A2.** The atom, like the solar system, is mostly **space.**

**A3.** **Yes,** if the substance in question is one of the basic elements.

**A4.** The electrons, or the flow of negative charges, must move **from the positive terminal of the battery through the battery to the negative terminal.**

**A5.** The nucleus contains **protons** (a proton has **one unit of positive charge)** and **neutrons** (a neutron has **one unit of positive and one unit of negative charge**). Electrons are in orbit around the nucleus. The **electron** is assigned **one unit of negative charge.**

## THE ION

If enough force or energy is applied to an atom it is possible to add or take away an electron or two. If this happens, the atom will have an unbalanced electrical charge (unequal number of electrons and protons). This unbalance will cause the atom to have either a negative or positive charge. When an atom is charged (either negative or positive), it becomes an **ion.** A negatively charged atom is called a **negative ion,** and a positively charged atom a **positive ion.**

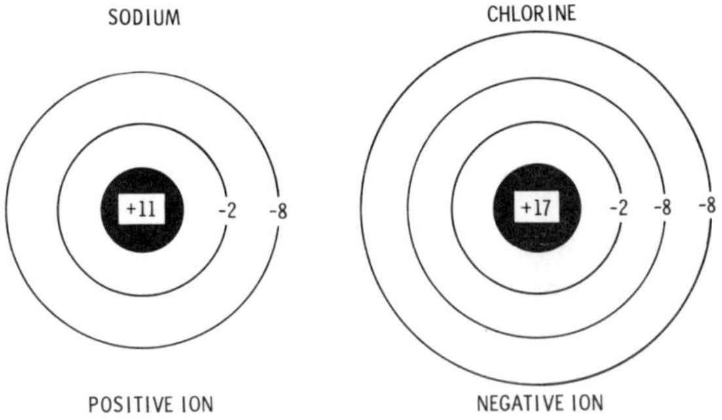


**Ions.**

It was shown earlier how the sodium and chlorine atoms were bonded to form common household salt. This bonding is not a difficult thing to understand if you consider that the sodium positive ion, having shared its outer electron with the now negative chlorine ion, so that the two bond together to form balanced outer-orbit zones. That is, the sodium atom shares its outer electron with the chlorine atom, which permits the chlorine outer-orbit zone and the sodium outer-orbit zone to be effectively filled.

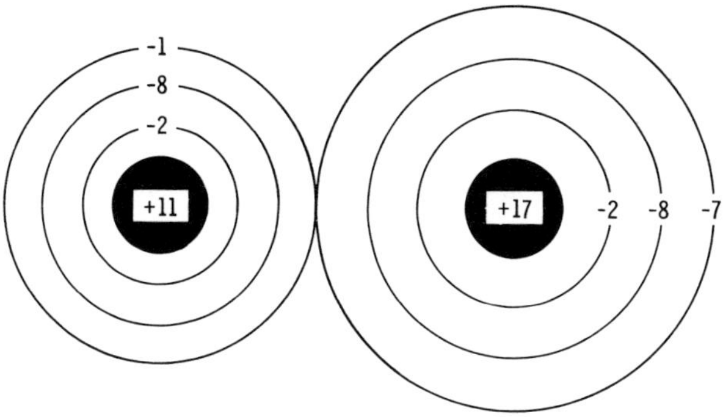


**Sodium and chlorine atoms.**

This sharing process establishes a stable combination (a molecule of salt) which is difficult to break. In the combination (molecule), you could consider the sodium atom capable of being a positive ion and, at the same time, the chlorine atom capable of being a negative ion if either should become isolated.

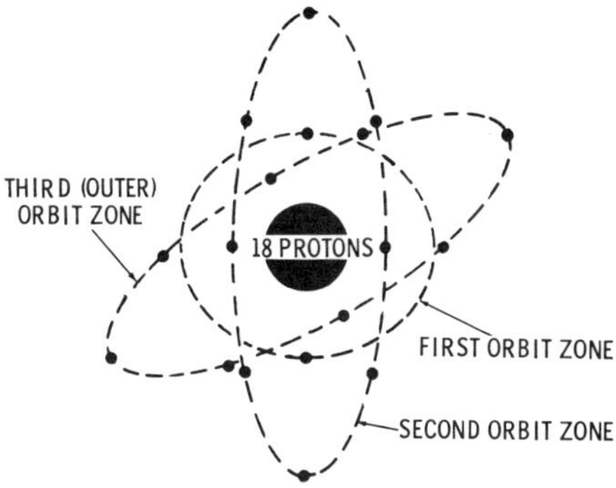


**The argon atom.**

**Q6. Could argon, with an atomic number of 18 (18 electrons and 18 protons), be considered a free-electron type element? Why?**

Your Answer Should Be:

**A6. No. Argon has all orbit zones completely filled.** Each argon atom has 2, 8, and 8 electrons in its three orbit zones, making it inert.

## FORCE AND FLOW

To have electrical movement, energy must be applied to the atoms in a material. For the electrical movement to be

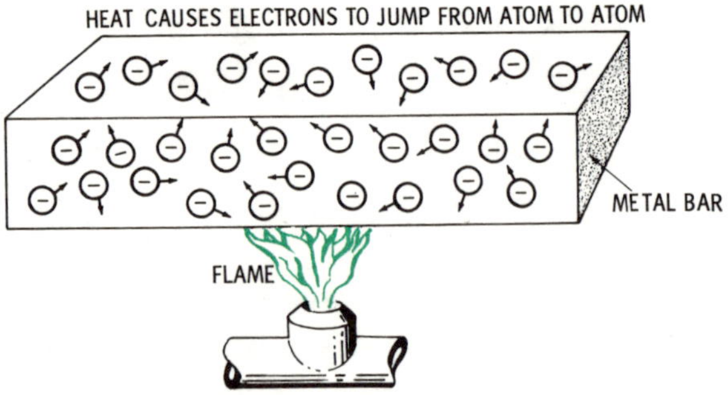


**Heat scatter effect.**

of value, the energy must be applied in such a manner as to move a relatively large number of electrons in the desired direction.

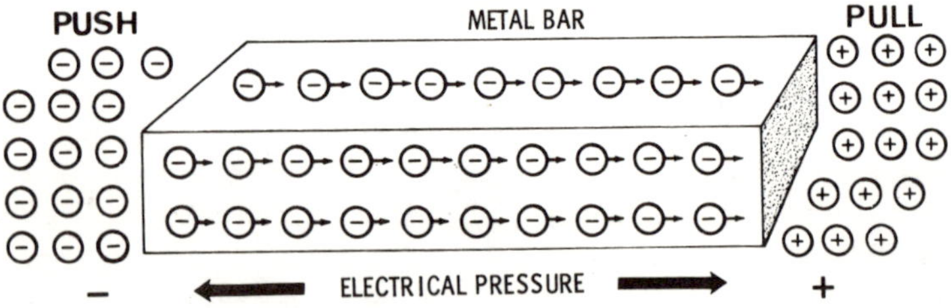

**Directed force.**

Electrical pressure is the most common form of control. This electrical pressure, or force, is made available by many forms of voltage sources.

A voltage source is an excess accumulation of negatively charged ions in one area with respect to another area.

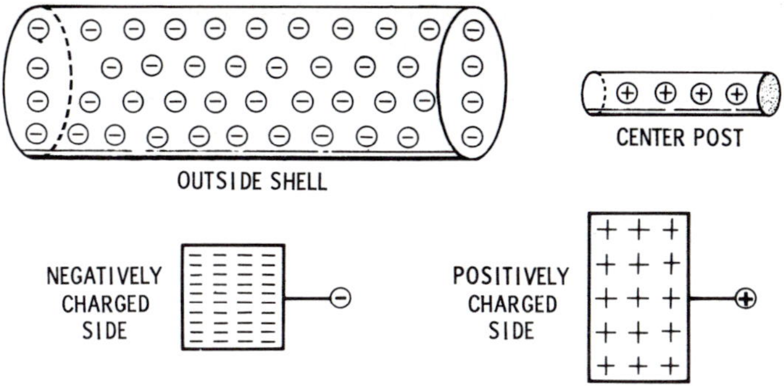


**A flashlight cell.**

A battery is one such form of voltage source. It has an excess of negative charges on one terminal and a lack of negative charges on the other terminal. When the two battery terminals are applied properly to other elements, a complete path from one terminal to the other is formed. The more negative terminal gives off negative charges that pass through the other elements to the positive terminal.

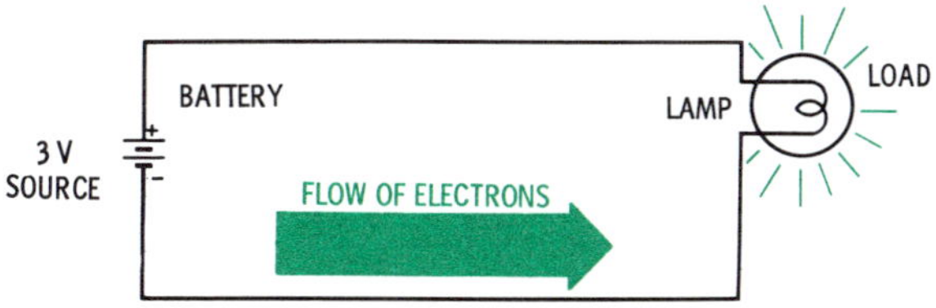


**Flow of electrons.**

This process continues until the charges on the two terminals are equal. This does not mean the terminals have to be so completely neutralized that no electrical charge exists on either terminal. Under this condition, the battery is discharged (run down).

Many processes in electricity are described in terms of electrical charge. Not only must you be concerned with the amount of electrical charge, you must also be concerned with the polarity (negative or positive).

**Q7. Consider the salt molecule. Could the effect of electrical charges, one atom to the other, hold it together?**

Your Answer Should Be:

**A7. Yes.** Electrical charges do affect the bonding of the two atoms, one to the other.

### Electrons and Ions

One atom is constantly trying to steal an electron from another atom so that the last orbit zone will be completely filled. This results in the production of two ions (if they are considered separately)—one negative and the other positive. The two atoms are thus held together because of the unlike charges attracting each other.

Why is it the earth does not go crashing into the sun? These two bodies are attracted to each other. If this attraction did not exist, the earth would not revolve around the sun in its yearly orbit. Instead, the earth would go rambling off into space. If you swing a ball in a circle at the end of a string, the ball tends to pull away from you. If you pull back with an equal force, the ball remains the same distance from you. The same is true of the earth and the sun. If the earth were to speed up (go faster in its orbit about the sun), it would slowly move farther and farther away from the sun. If it were to slow down, it would finally collide with the sun.

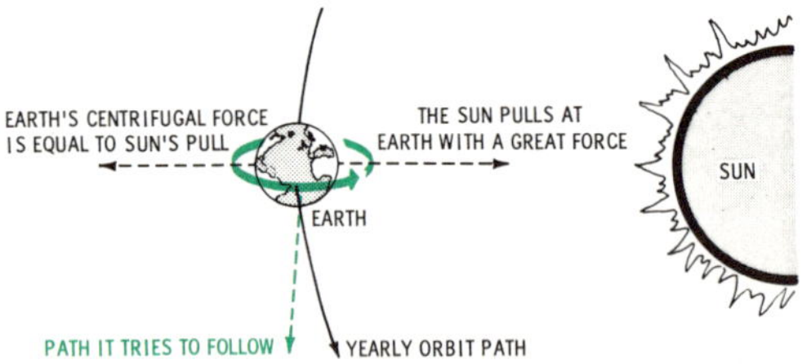


**Force and the sun.**

With respect to the ratios of size and weight indicated between the earth and the sun, how do the particles in the tiny atom compare? First, the electron is many times lighter than the proton. Second, the proton is slightly lighter than the neutron. With nearly all the mass or weight in the center (nucleus) of an atom and the lighter particle moving about it in an orbit, the electron is the most likely particle to be disturbed by some external force. However, the entire

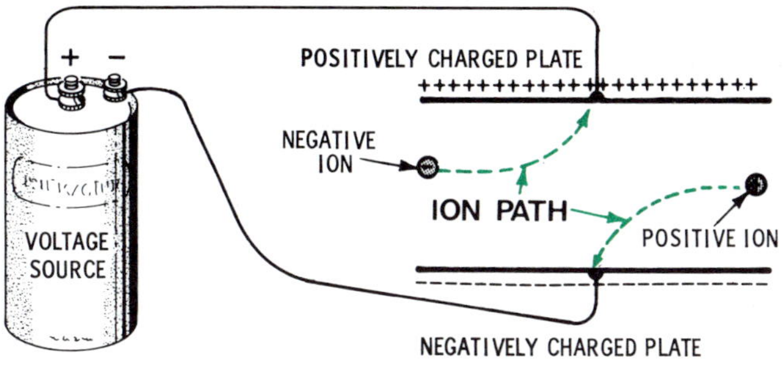


**The ion path.**

atom may be moved. How about the ion, could it be caused to travel in a directed line? Yes. If positive, it will be repelled by another positive force or attracted by a negative force.

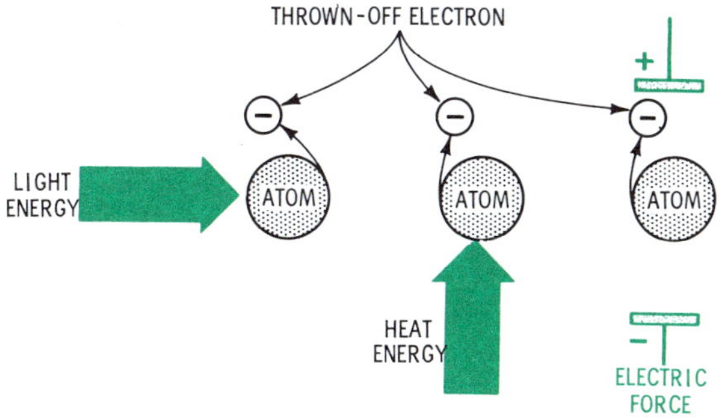


**Heat, light, and electricity affect the electron.**

**Q8. What happens in the atom, with reference to the description of forces on the earth? Draw the atom to include the nucleus and the orbiting electrons.**

**Q9. Could the earth be made to move in a straight line?**

**Q10. Show by an illustration what effects positive and negative forces have on a negative ion.**

**Q11. Draw a lithium atom, atomic number 3. Label the free electron.**

**Q12. What effect do the neutrons within an ion have on the movement of the ion?**

**Your Answers Should Be:**

**A8.**

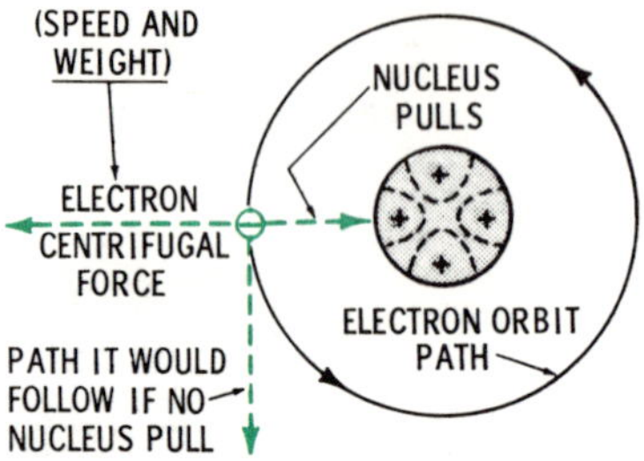


**A9. Yes,** if enough force were applied from some source that had a much greater pull than that applied by the sun.

**A10.**

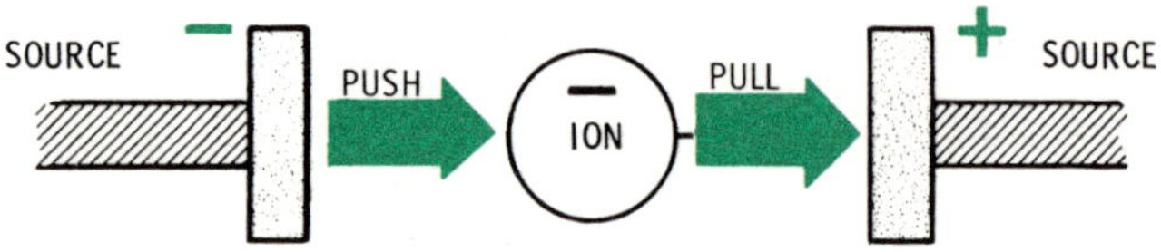


**A11.**

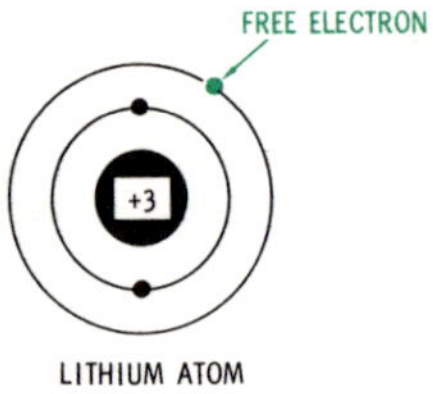


**A12.** They have the effect of **slowing down the movement of the ion** because of their mass. Without the neutron the ion could be moved more easily.

## PRODUCTION OF ELECTRICITY

Electricity is produced by a movement of charges between two terminals. One terminal collects negative charges and the other positive charges. Electricity can be the movement of free electrons from one point to another. A DC generator, a device for producing a constant voltage, moves an excess of electrons in one direction through a series of windings. The electron movement is caused by the loops of wire cutting magnetic fields.

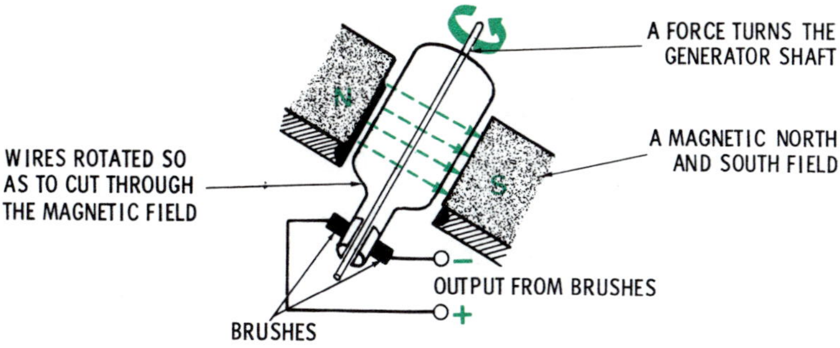


**A generator.**

Electricity can also be a movement of positive ions in one direction and, at the same time, a movement of negative ions in the other direction.

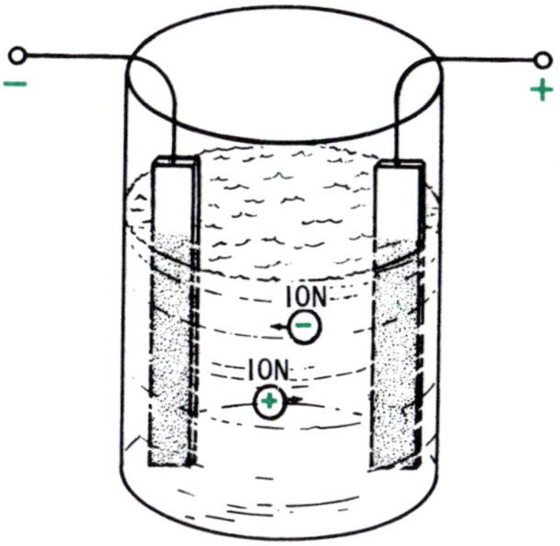


**Internal action of a battery.**

In order for electricity to be produced, there must be a way to transfer charges from one place to another. This transfer of charges is made possible by the use of **conductors.**

**Q13. What is the source of electricity called?**

**Q14. How is the source of electricity made available to other locations?**

**Your Answers Should Be:**

**A13. A voltage source.**

**A14.** Electric charges are transferred from one location to another by **conductors.**

## CONDUCTORS AND INSULATORS

The use of a conductor as the path by which electrical charges are transferred from one point to another is demonstrated in automobiles, homes, aircraft, ships, and almost anything else you can name. Examples of insulators are the large glass shields used to support wires and the rubber or plastic coverings along the surface of the conductors.

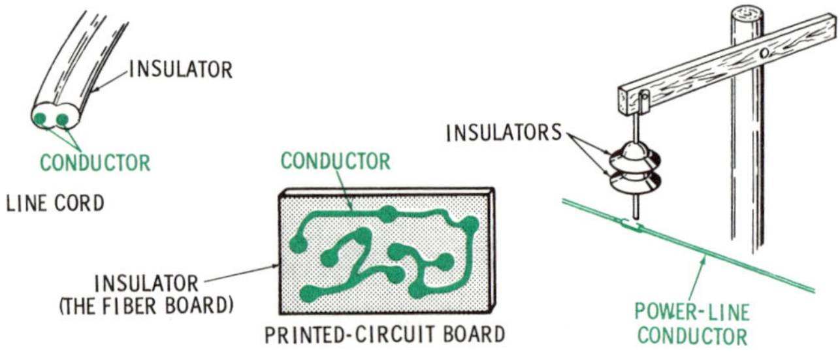


**Conductors and insulators.**

### Conductors

A conductor is a free-electron material which serves as a path for electric current.

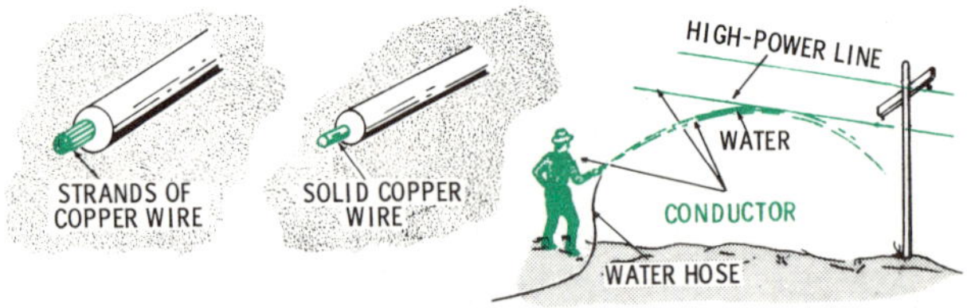


**Conductors.**

Some materials are better conductors of electricity than others. The availability of more free electrons in silver and copper, for example, makes them better conductors than iron or steel.

## Insulators

An **insulator** is a material through which electric current will not flow easily; thus, an insulator may be used to separate conductors from each other. An insulator may also be a covering over each conductor, permitting two or more conductors to be positioned near each other without danger of current following an undesired path between the conductors. In all cases, the insulator, to be effective, must present a path of high opposition to electron or ion flow.

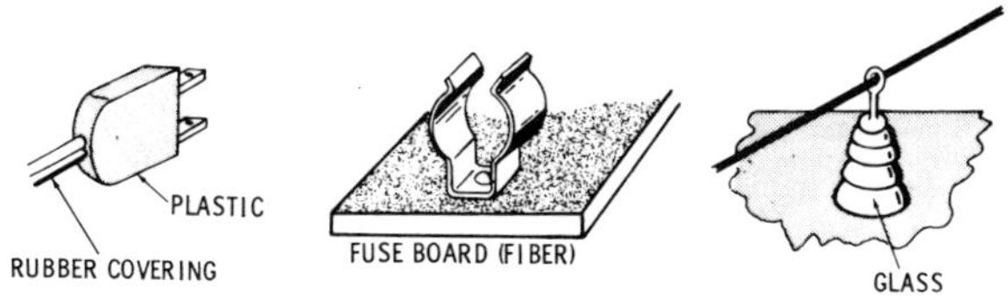


**Insulators.**

Any insulator, however, will become a conductor of electric current if subjected to a sufficiently high voltage. This is true because of the atomic structure of all elements. The insulators in the following list will remain insulators under normal values of house voltage (120 volts).

| Conductors | Insulators |
|---|---|
| silver | glass |
| gold | porcelain |
| copper | plastic |
| lead | rubber |
| tin | mica |
| brass | ice and snow (pure) |
| aluminum | nylon |
| bronze | bakelite |
| nickel | paper |
| iron and steel | wood |
| cadmium | paraffin |
| graphite | quartz |
| mercury | dry air |

**Q15. With respect to the orbit zones, how should an atom in an insulator be constructed?**

**Q16. With respect to the orbit zones, how should an atom in a conductor be constructed?**

**Your Answers Should Be:**

**A15.** Each atom should have **completely filled** orbit zones.

**A16.** Each atom should be a **free-electron type element.**

## STATIC ELECTRICITY

**Static electricity** is the production of a difference in electrical charges between two materials as a result of friction. Rubbing one type of material against another, each of different orbit-zone construction, causes electrically charged particles to be transferred from one material to the other.

**Charging a balloon.**

Lightning is an example of static electricity, and is the result of an accumulation of charges caused by friction between cloud layers or between clouds and the earth. When the charge becomes great enough, it breaks down the insulating air. The resulting discharge produces the familiar lightning flash.

The sparks that jump from your fingers to metal objects or to other persons is another example of static electricity. The sparks are the result of your body becoming charged by walking across a rug and discharging to an object having a neutral or opposite charge.

**Charging a cloud.**

## Electrical Charge

Electrical charge is founded on a basic electrical principle. The **electron** is considered to be a **negative particle,** and the **proton** is considered to be a **positive particle.** These are the two elementary electrical particles on which all other expressions of electrical charge are based.

An ion is another example of a negative or positive charge. Whether the ion is positive or negative depends on the number of electrons in the outer orbit and the number of protons in the nucleus. If the electrons outnumber the protons, the atom is a negative ion. If the protons outnumber the electrons, the atom is a positive ion. It is very difficult to determine if a single atom is an ion, and whether it is positive or negative even after it has been found to be an ion.

There are ways to measure the overall effect of an accumulated charge. The most common method is with a meter, such as a voltmeter, ammeter, or a combination unit known as a multimeter.

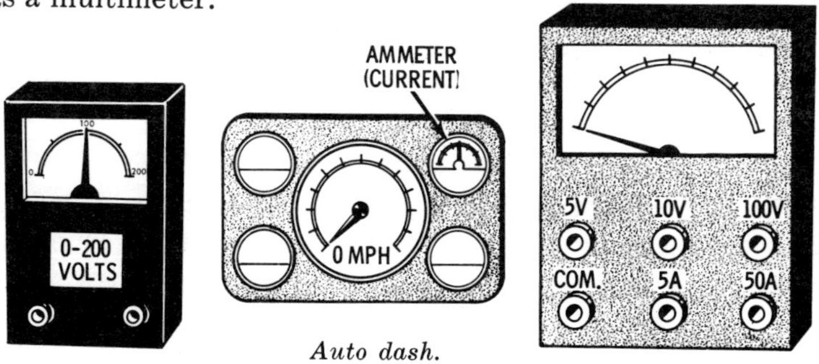


*Voltmeter.* *Auto dash.* *Multimeter.*

**Meters.**

These devices indicate an approximate measurement of charges. Yet, these measurements are sufficiently accurate to serve as standards on which electrical and electronics personnel base their evaluations.

**Q17. The electron is a --------- particle.**

**Q18. The proton is a --------- particle.**

**Q19. What two factors determine whether an ion will be charged positively or negatively?**

**Q20. How could you measure the overall effect of an accumulated charge?**

**Your Answers Should Be:**

**A17.** The electron is a **negative** particle.

**A18.** The proton is a **positive** particle.

**A19.** **The number of electrons in the outer orbit and the difference in number between electrons and protons in the atom.**

**A20.** You can measure the overall effect of a charge by the use of **a meter.**

## Like and Unlike Charges

Another principle on which all other concepts of electricity are founded is the fact that **like charges repel** and **unlike charges attract.**

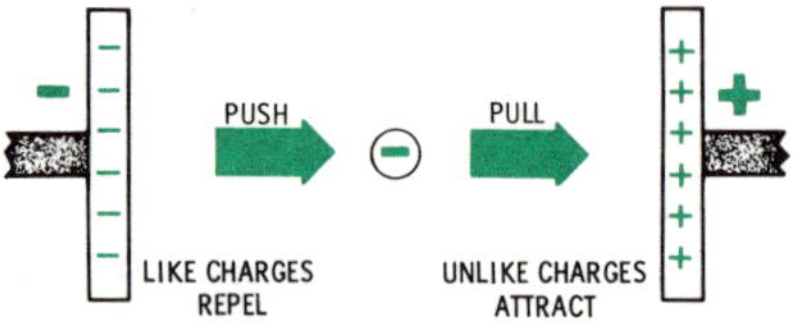


**Charges.**

A negatively charged body placed near a positively charged body produces a force of attraction between them. One body tries to contact the other in an attempt to neutralize, or balance, the charges. You can try a simple experiment to demonstrate this principle. Inflate a rubber balloon and rub it on your hair. (Make certain your hair is dry and clean.)

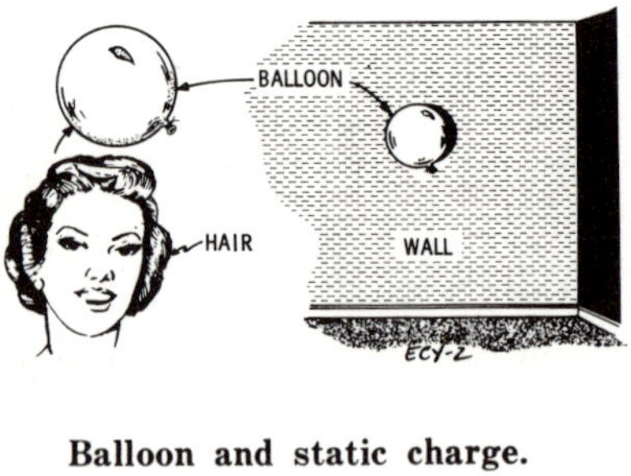


**Balloon and static charge.**

Now place the balloon near a wall. The balloon clings to the wall because of the negative static charge it accumulated.

When a charged particle comes in contact with a neutral particle, the charge will be equalized (divided equally) between the two particles. If the original charge is great enough, both particles will then acquire a like charge and repel one another. This can easily be seen if a pith ball is given a negative charge and allowed to come in contact with a neutral pith ball.

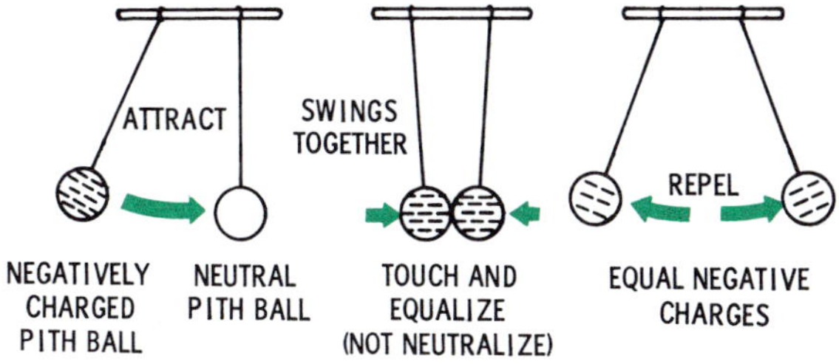


You can perform another experiment using an electroscope to illustrate the principle of attraction and repulsion between static charges.

What you will need to build an electroscope:

1. One clear glass bottle (1 pint) and a rubber cork to fit.
2. One metal rod, approximately ¼ inch in diameter.
3. Two thin pieces of aluminum foil (1 inch × ¼ inch).

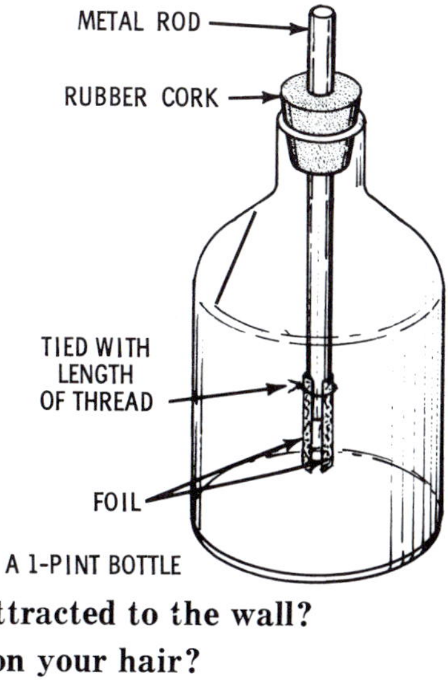


An electroscope.

**Q21. Why was the balloon attracted to the wall?**

**Q22. What charge was left on your hair?**

**Your Answers Should Be:**

**A21.** Because **the wall has a neutral charge** and **unlike charges attract** one another.

**A22. A positive charge.**

**Electroscope Demonstration**

Testing the electroscope:

1. Rub a rubber comb with fur.
2. Touch the metal rod with your finger.
3. Touch the metal rod with the charged comb.
4. The pieces of foil should repel one another.

Next test:

1. Touch the metal rod with your finger.
2. Charge the comb using the fur.
3. Bring the comb near the rod (do not touch the rod). Now touch the rod with a finger from your other hand.
4. Remove your finger from the rod.
5. Remove the comb from the vicinity of the metal rod.
6. The foil strips should repel one another.

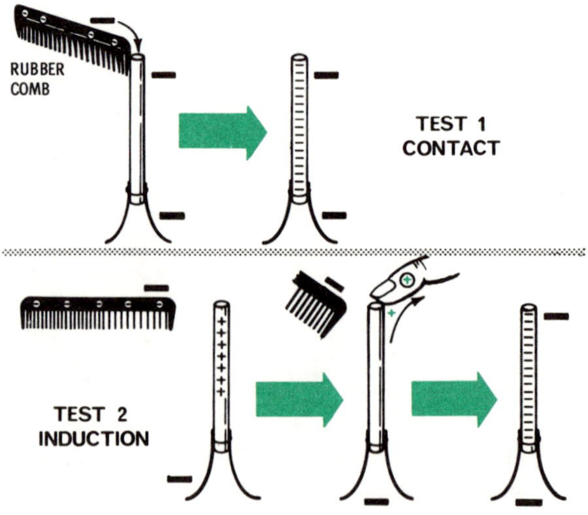


**Charging by contact and induction.**

## Forces Between Charged Objects

You can visualize the effects of charged bodies on one another and on neutral bodies. First, what is the effect called which caused the foils to be repelled, or the pith ball to be attracted or repelled? The answer is **force.** How do you determine just how much force and what caused it to exist? The answer is **charge.** The amount of charge determines the amount of force. How can the charge be measured? Each object must be measured separately with respect to a neutral object.

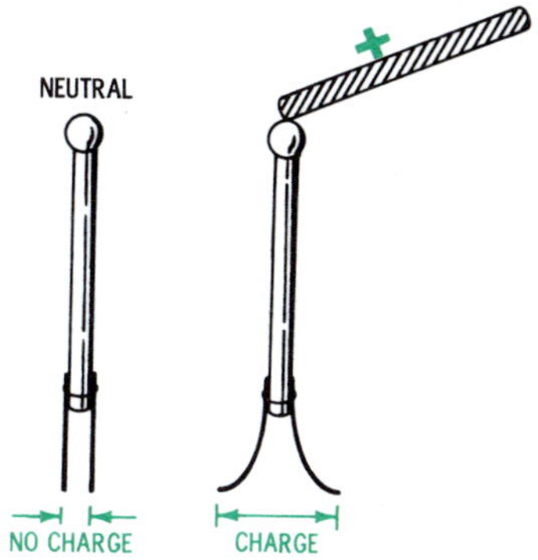


**Measuring charge.**

Multiply the numerical value of charge on one object by the numerical value of charge on the other. The product gives the total charge. The greater the total charge, the greater is the force.

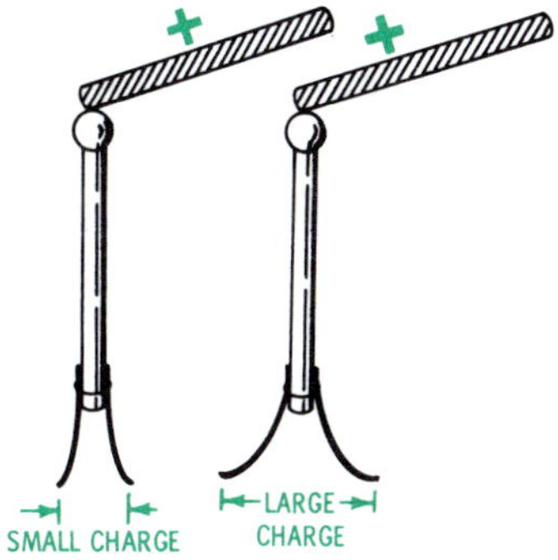


**More charge—more force.**

**Q23. What is the charge condition on the metal rod and the comb at the completion of the first test?**

**Q24. What is the charge condition on the metal rod and the comb at the completion of the second test?**

**Your Answers Should Be:**

**A23.** **Equal and negative,** because the comb took electrons from the fur and transferred them to the metal rod.

**A24.** **The metal rod is negative** because the comb repelled electrons away from the tip of the rod and the excess positive charges were removed via the finger. The comb retains a **negative charge.**

Does the distance between the objects have any effect on the force between them? Yes. The greater the distance, the less effect one body will have on the other.

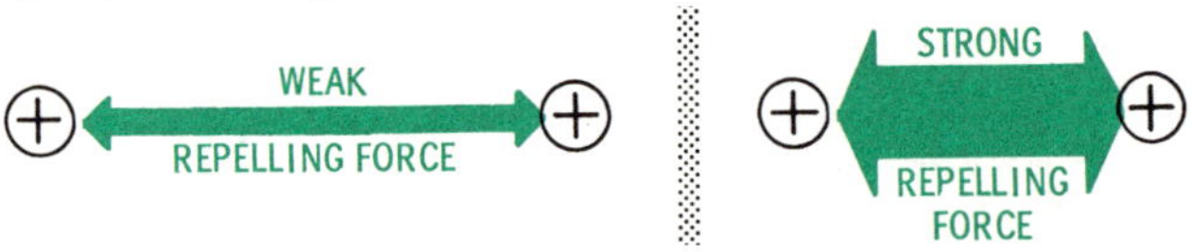


**Less distance—more force.**

There is one very important concept you must understand before the mathematical expression which fits all of these ideas can be determined. This concept states that the distance is not a direct measurement factor. For instance, if the distance is doubled between the two objects, the force effect between them will not be half as great, as might be expected, but only a fourth as great. Why is this?

Consider the effect that only one charge has on the other. If this charge is moved twice the distance away from the other, it will have one half the effect as before. By the same reason, the other charge will have one half the effect as before. Therefore, one half times one half is one fourth.

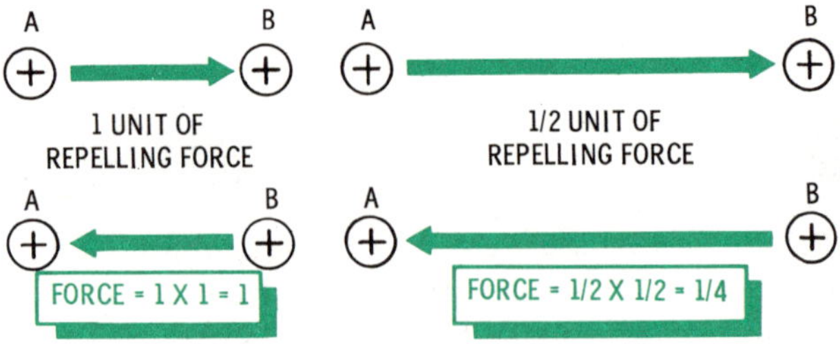


**The effect of distance on force.**

Now that the reasoning has been established, take a look at the shorthand statement in a mathematical form.

$$F = \frac{Q_1 \text{ times } Q_2}{d \text{ times } d} = \frac{Q_1 \times Q_2}{d^2}$$

where,

> F is the force exerted between two charged bodies,
> $Q_1$ is the charge on the first body,
> $Q_2$ is the charge on the second body,
> d is the distance between the two bodies.

The force will be in dynes, if the Q's are given in electrostatic units and the distance is measured in centimeters (cm). An **electrostatic unit** is defined as the charge required to repel an equal charge 1 centimeter away, with a force of 1 unit. The unit of force is the **dyne.**

What is a dyne? First, what is force? One way of reasoning is to think of the effect you have on the speed of a toy wagon by pushing it. It requires more effort on your part to get it started. After it starts, you need only push with a small effort to keep it at a constant speed. What if the push against the wagon is always just as much as that with which you started? The speed of the wagon will increase and the force of your push must increase accordingly. The speed of the wagon and the force of your push will increase until the limit is reached where you cannot push any harder. At this point the force against the wagon decreases and the speed will no longer increase. If the pushing force decreases below the amount required for a steady speed, the wagon will slow down.

Solve the following problems, using the force expression.

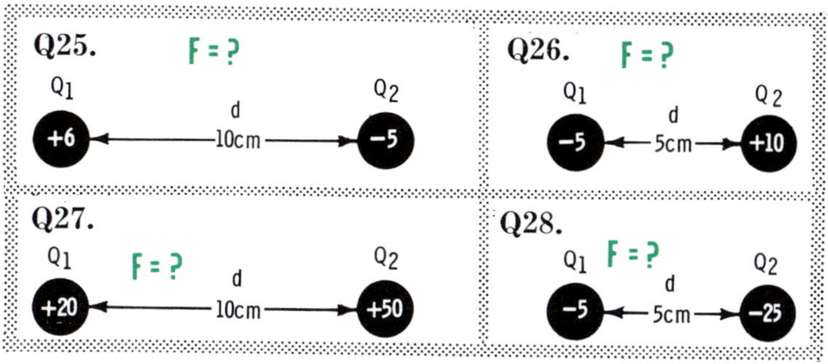

**Your Answers Should Be:**

A25. F = (6 × 5) ÷ 100 = **0.3-dyne attraction.**

A26. F = (5 × 10) ÷ 25 = **2-dyne attraction.**

A27. F = (20 × 50) ÷ 100 = **10-dyne repulsion.**

A28. F = (5 × 25) ÷ 25 = **5-dyne repulsion.**

### Application of Force

A dyne is the force required to cause a 1-gram mass to travel a distance of ½ centimeter (cm) when the force has been applied for 1 second. If this same 1 dyne of force is continued for another second, the gram of mass will be 2 cm from where it started. At the end of the first second it will be traveling at a speed of 1 cm per sec. At the end of the second second it will be traveling at a speed of 2 cm per sec.

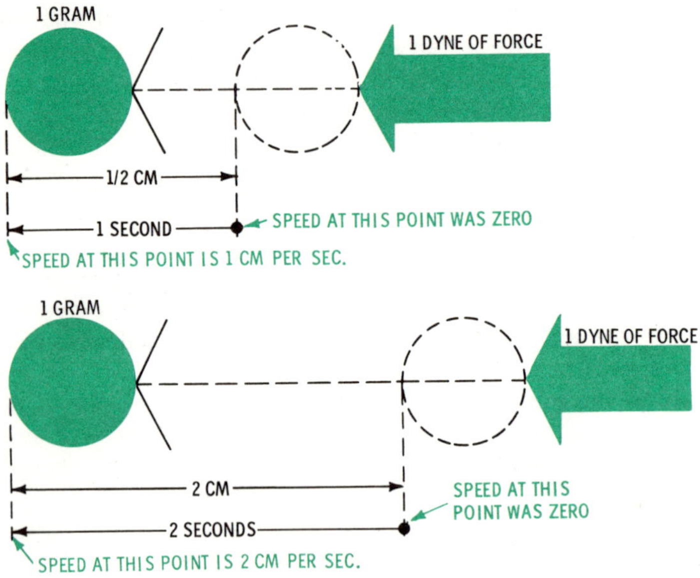


**Force-mass-speed.**

A force of one dyne produces an **acceleration** of 1 centimeter per second per second on a 1-gram mass.

One gram of matter is the measurement of how much resistance a collection of matter offers to a change of motion. One dyne of force applied to a 1-gram mass results in the following speeds and distances traveled:

| | | | | | | | | |
|---|---|---|---|---|---|---|---|---|
| **TIME** seconds | 0 | 1 | 2 | 3 | 4 | 5 | 6 | 7 |
| **SPEED** cm per sec. | 0 | 1 | 2 | 3 | 4 | 5 | 6 | 7 |
| **DISTANCE** centimeters | 0 | ½ | 2 | 4½ | 8 | 12½ | 18 | 24½ |

## Graphical Representation of Forces

There are many methods used to represent the force between two charged bodies. These methods attempt to make the idea of force between the two bodies easier to understand or visualize. The following diagram is not one of the classical representations. Instead it emphasizes the potential difference from one plate to the other. Notice that the force is equally distributed throughout the entire area between the two plates.

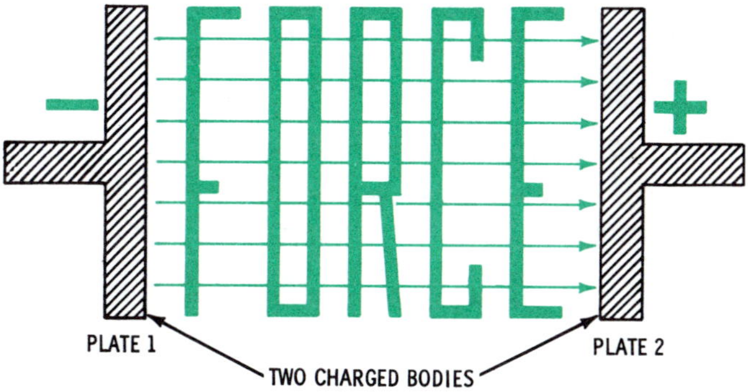


**Force exerted by two charged bodies.**

**Q29. What is the electrostatic unit of force?**

**Q30. One ____ is the force required to cause a one ____ mass to travel a distance of ½ __ when the force has been applied for one ______.**

**Your Answers Should Be:**

**A29.** The **dyne.**

**A30.** A **dyne** is the force required to cause a one-**gram** mass to travel a distance of ½ **cm** when the force has been applied for one **second.**

## Electric Fields

In addition, there are other potential differences existing between imaginary points located somewhere between the two charged plates.

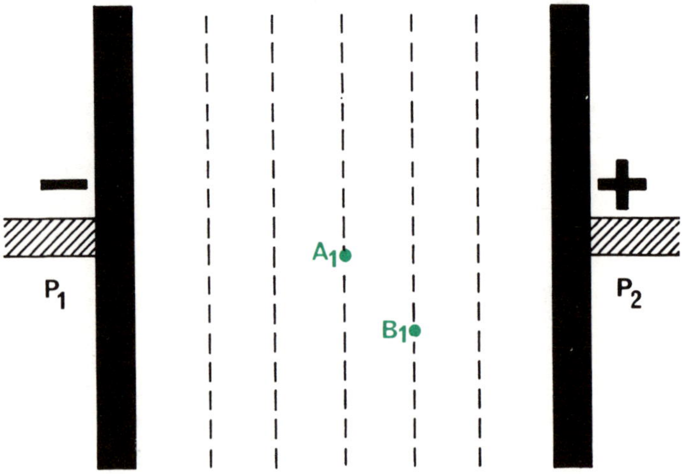


**Potential difference between charged plates.**

Because of the distance between $P_1$ and $P_2$ and the amount of charge on each, there is a specific force between the two plates. The force in this case is an electrical attraction of one plate to the other.

If a charged particle is placed at point $A_1$, a force acts on the particle. For example, an electron placed at point $A_1$ would be urged toward plate $P_2$. Similarly, a force acts on a charged particle placed at point $B_1$ or any other point between the two charged plates. In the case of two parallel charged plates, the amount of force acting on a charged particle is the same for any position of the particle between the plates.

The region around the charged body is referred to as the **electrostatic field of force.** The lines represent the path a free electron would follow.

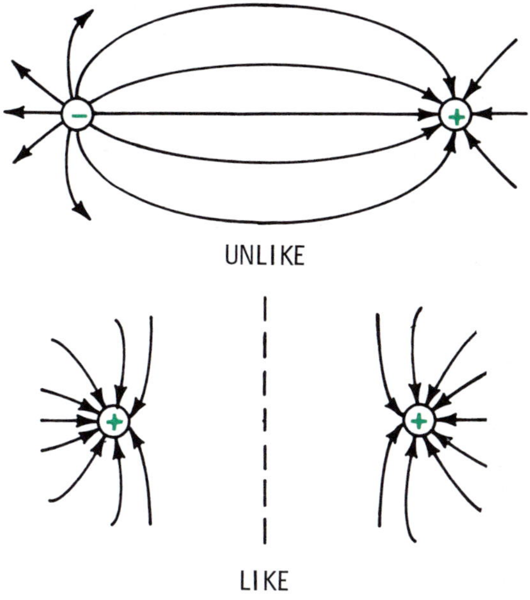


**Electrostatic fields.**

This area therefore contains electrostatic lines of force between two charged bodies. There is an electrostatic field around each charged body. When two positively or two negatively charged particles are placed near each other, their electrostatic fields repel each other. However, when a positively charged particle is placed near a negatively charged particle, their electrostatic fields attract each other. **Like charges repel,** and **unlike charges attract.**

**Q31. What two factors must be considered in determining the force exerted between two charged bodies?**

**Q32. The area surrounding a charged body is called a(an) ______________ _____ __ _____.**

**Q33. Like charges _____, and unlike charges _______.**

**Your Answers Should Be:**

**A31.** The **distance** and the **potential difference** between the two charged bodies.

**A32.** The area surrounding a charged body is called an **electrostatic field of force.**

**A33.** Like charges **repel,** and unlike charges **attract.**

## ELECTRIC CURRENT

Electric current is the movement of electrical charges from one location to another. This is normally considered as the flow of current (electrons) through a conductor. It can also be the movement of charged particles through a battery or any other electrical component.

Even lightning is the result of the movement of charges from one location to another. It is ion movement, primarily, with positive ions moving in one direction at the same time negative ions are moving in the opposite direction. The movement is very fast and involves a tremendous number of ions. This results in a very intense light which is the result of a high concentration of forces discharging from one cloud to another, or from a cloud to the earth, and causing a large number of ions to flow.

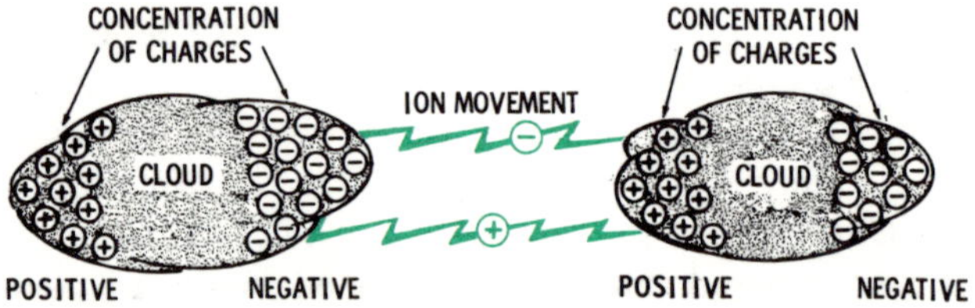


**Lightning.**

### Moving Charges

Charges in motion, then, are actually the movements of either free electrons or ions from one location to another. This constitutes current flow, or **electric current.** Some people describe the electrical flow of charges in terms of the electron theory and others describe electrical flow in terms of ions. Those that propose current to be a flow of electrons employ the **electron-current theory.** When current is considered to be positive ions, the theory explaining its flow is called **conventional current flow.**

Current is defined as the **movement of electrons through a conductor.** Current is also the **movement of ions through a material.** The most important factor you must remember from discussions on the subject is the application of **electron** and **conventional** current theories.

### Electron Current Theory

Electron current is said to flow through a circuit when electrons are repelled by the negative terminal of a voltage source and are attracted by the positive terminal. Inside the source, electrons travel from positive to negative.

**Electron current flow.**

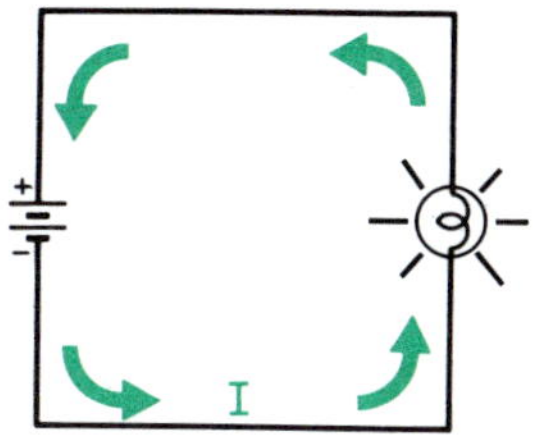


### Conventional Current Theory

When current is explained as leaving the positive terminal of a source and flowing through a circuit to the negative terminal, the **conventional current-flow theory** is being used.

**Conventional current flow.**

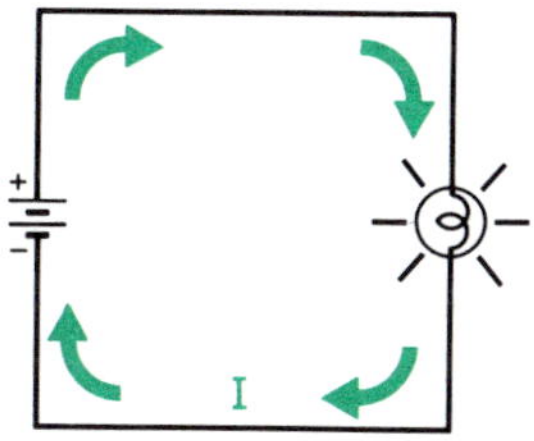


**Q34. Why can't electron current be the only theory employed?**

**Q35. Will the movement of negative ions produce the same electrical effect as the movement of electrons?**

**Q36. Can electrons be in motion from atom to atom in the opposite direction to the motion of positive ions?**

**Q37. Can both negative and positive ions be in motion when current is said to flow?**

**Your Answers Should Be:**

**A34.** Because the movement of electrons **can create ions which will also move.**

**A35. Yes.** The negative ions will be caused to move by the same force that moves the electrons.

**A36. Yes.** The same force which causes positive ion motion in one direction has exactly the opposite effect on the electron.

**A37. Yes.** This will be clearly demonstrated later in this chapter.

## Current Units and Symbols

Just as there are measurement units for height, weight, time, force, etc., there is also a measurement unit for current. If this were not true, you could not measure current, determine current from other known factors, or describe a quantity of current to others. The unit of measurement for current is called an **ampere.**

An ampere is the effect of 6,250,000,000,000,000,000 electrons passing any point in an electrical circuit in 1 second. It is easy to remember this number as 6.25 million, million, million electrons, which can be written as 6.25 MMMe. It is difficult to use such a figure continuously, so the term **coulomb** is used in its place. That is, a coulomb is 6.25 million, million, million electrons.

To measure electric current in a circuit, you must measure how many charged particles pass any point in the circuit per unit of time. The shorthand symbol assigned to represent current is **I.**

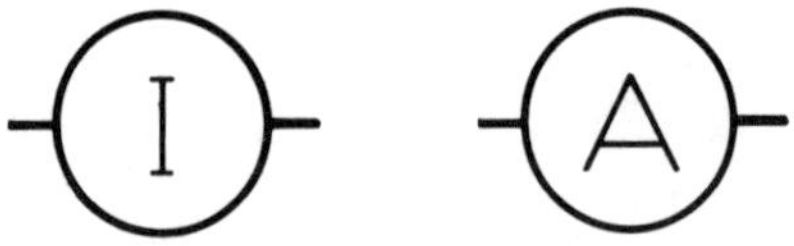


**Current-meter symbols.**

An ampere is the standard unit of measure for current (I) in a circuit. However, there are methods for describing very small or very large currents.

**Magnitude or Strength of Current**

Feet and inches are smaller units than yards, a measurement of length. The measurement unit for current, an ampere, can likewise be subdivided. A **milliampere,** for example, is 1/1000 (one thousandth) of an ampere. A **microampere** is 1/1,000,000 (one millionth) of an ampere. **Amp, milliamp,** and **microamp** are accepted abbreviations of ampere, milliampere, and microampere.

An ammeter often employs the smaller units of current measurement. Meters must be capable of measuring different values of current and, in some cases, revealing the direction of current flow.

An ammeter is a device having scales calibrated in amperes, milliamperes, or microamperes, used for measuring the total current flow in a circuit.

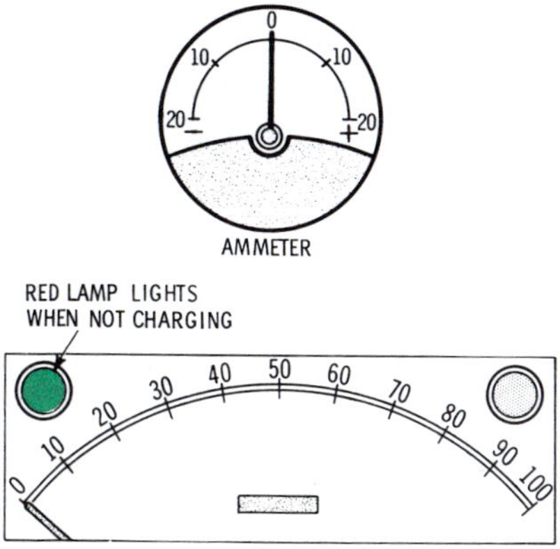


**Auto ampere indicators.**

**Q38. The unit of measure for current is the ______.**

**Q39. A(an) ________ is 6.25 MMM electrons.**

**Q40. The shorthand symbol for current is ____.**

**Q41. How can you measure the current in a circuit?**

**Q42. What is an ammeter?**

**Q43. How should an ammeter be connected to a circuit?**

**Your Answers Should Be:**

**A38.** The unit of measure for current is the **ampere.**

**A39.** A **coulomb** is 6.25 MMM electrons.

**A40.** The shorthand symbol for current is **I.**

**A41.** You can measure the current in a circuit with an **ammeter.**

**A42.** An ammeter is an electrical device, having scales calibrated in amperes or portions of an ampere, placed in an electrical circuit in such a manner as to indicate how much current is flowing.

**A43.** It must be placed in the circuit so as to monitor all of the current in the circuit.

**Current Measurement**

An ammeter in an automobile indicates both minus and plus charges (amperes). If the meter indicates minus, the auto is using electrical power from the battery. If the meter

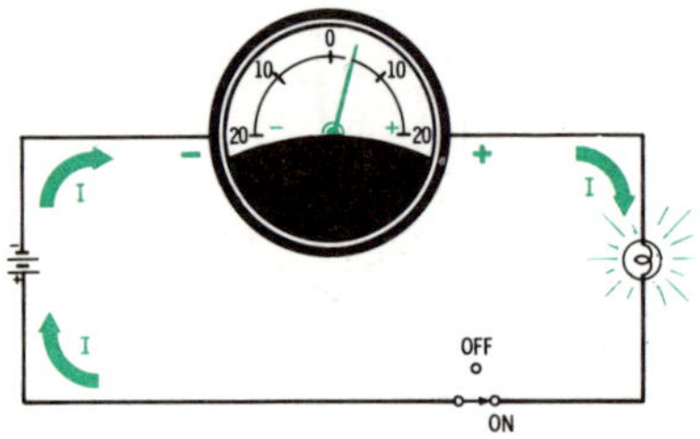


**Ammeter in a circuit.**

indicates plus, the auto is getting electrical power from the generator and, at the same time, the battery is being charged. Therefore, a minus meter reading indicates the battery is being used, and a positive meter reading indicates the battery is being charged. Before the days of long trips and dependable voltage regulators on the automobile, the operator had to watch the ammeter very closely.

The reason for the close observation was not only to insure a well-charged battery for starting purposes, but also to prevent an overcharge of the battery. Overcharge could damage the battery, and undercharge could require manual cranking or a push to start the engine. Even close observation was not entirely satisfactory because it was a hit-and-miss arrangement, and did not give the accuracy required for long life and good dependability. Thus, the voltage regulator was developed, and man did not have to know quite so much to operate his "horseless carriage." The automobile mechanic, however, had to know more and more about electricity.

The requirement for electrical knowledge by a mechanic is even greater today. Electrical devices in the automobile are becoming more complex. This means a mechanic must be better qualified in the use of electrical tools, electrical systems, and the fundamental knowledge of electricity so that he can identify maladjustments, failures, and any number of other problems which need his attention if the automobile is to perform properly. The use of an ammeter is important to many others, for the automobile is only one application of electrical devices.

Ampere symbols and their meanings are shown below.

| Current Designation | Meaning |
|---|---|
| I | symbol for current |
| 1 ampere (amp) | 1 coulomb per second |
| 1 milliampere (ma) | $\frac{1}{1,000}$ ampere = 0.001 amp |
| 1 microampere ($\mu$a) | $\frac{1}{1,000,000}$ ampere = 0.000001 amp |

**Q44. How much current is indicated on the following meters?**

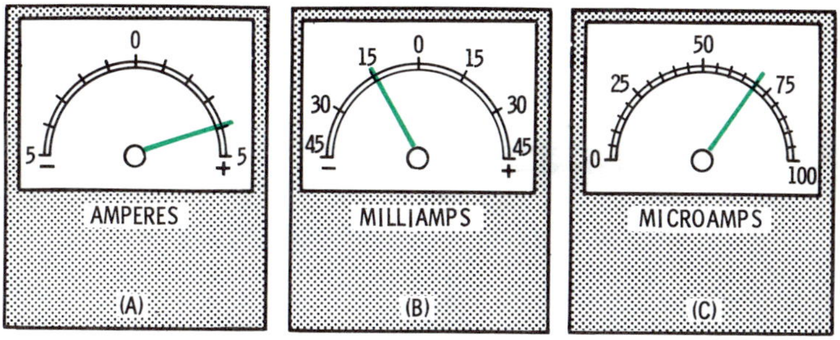

**Your Answers Should Be:**

**A44.** (a) **+4 amps** (b) **—15 ma** (c) **+70 $\mu$a**

## VOLTAGE

Electrical pressure required to move current through a circuit is called voltage. It is the accumulation of negative electrical particles on one terminal with respect to positive electrical particles on the other terminal of a voltage source. It is the force that causes the movement of electrical particles through a circuit. Voltage, as a pressure, is often called **electromotive force.**

### Electromotive Force

The abbreviation for electromotive force is **emf.** This is the force which causes electricity to flow when there is a difference in electrical charge between two terminals.

### Potential

**Potential** is another term associated with electrical pressure, emf, and voltage. All of these terms are used interchangeably. The term potential generally denotes, however, that voltage (emf or pressure) is available but not necessarily being used to cause current flow. Voltage may be described as a **potential drop, potential difference,** or **voltage potential.**

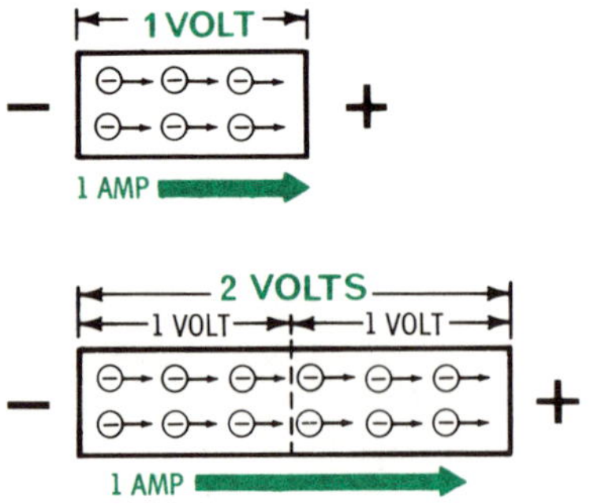


**Voltage potential.**

You can see that current will be the same through a conductor (with a specific voltage applied) as it is through a similar conductor twice as long but with twice the voltage applied. In each case, you could describe the voltage applied as a potential difference between the two ends of the conductor. What happens if the conductor is twice as long and the same potential difference is applied?

As in the previous illustration, the conductors are of the same material and cross-sectional area. The second conductor is twice the length of the first and thus offers twice as much material for current to flow through. Twice the length of material means that current will encounter twice as much opposition.

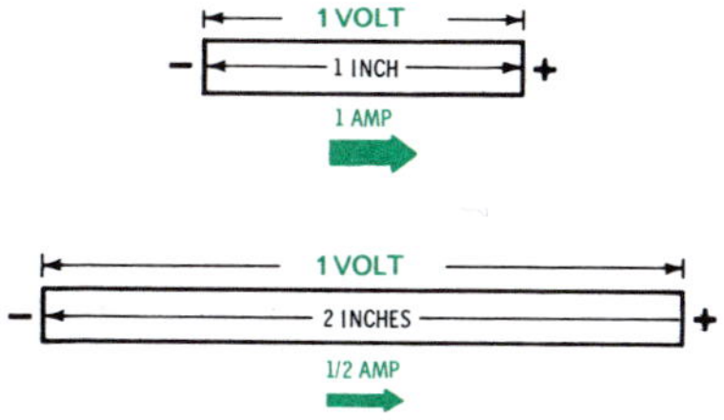


**Voltage division.**

## Voltage Definition

**Voltage** is **electrical pressure.** It is the force which causes current to flow. Voltage can be described as the **potential difference** between any two points.

## Voltage Units and Symbols

The standard unit of measurement for electrical pressure is the **volt.** A volt is that amount of pressure which causes one amp to flow through a resistance of one ohm. The volt may also be described in smaller or larger units. Microvolts are used to define small voltages.

**E** is the letter symbol for voltage, and is used to represent pressure in volts when working with electrical circuits.

**Q45. Electrical pressure is called ________.**

**Q46. The unit of measure for electromotive force (emf) is the ____.**

**Q47. What is voltage?**

**Q48. When 1 volt is applied across a 1-inch conductor, 1 amp of current is produced. How much current will flow if 1 volt is applied across a 2-inch conductor of the same material?**

**Your Answers Should Be:**

**A45.** Electrical pressure is called **voltage.**

**A46.** The unit of measure for electromotive force (emf) is the **volt.**

**A47.** Voltage is **the force that causes current to flow.**

**A48.** Only **0.5 amp** of current will flow.

**Comparison of Voltage and Current**

There are many values of voltage and current. For example, voltages in a flashlight range from 1.5 to 9 volts or higher; corresponding current in flashlights varies from a few ma to 2 amps. Voltage in an automobile may be 6 or 12 volts; current varies from a few ma for small lamps to several hundred amps for the starter. Voltages in a home vary from 16 to 240 volts, depending on where you live and what electrical devices are installed; and currents range from a few ma to 20 amps or more.

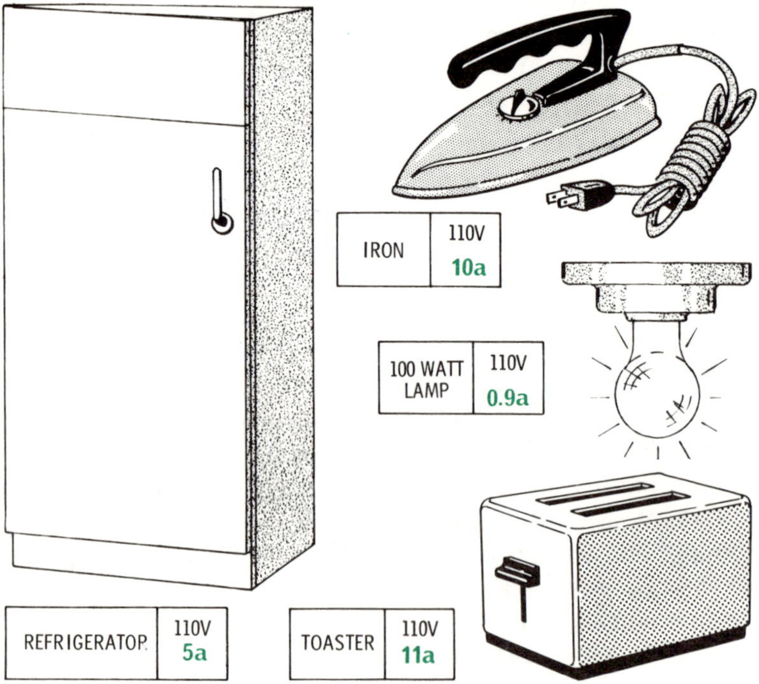

## CHEMICAL VOLTAGE SOURCES

There are many ways to produce electricity, but there are only six fundamental methods used to change another form of energy into electrical energy. Five of them are shown below.

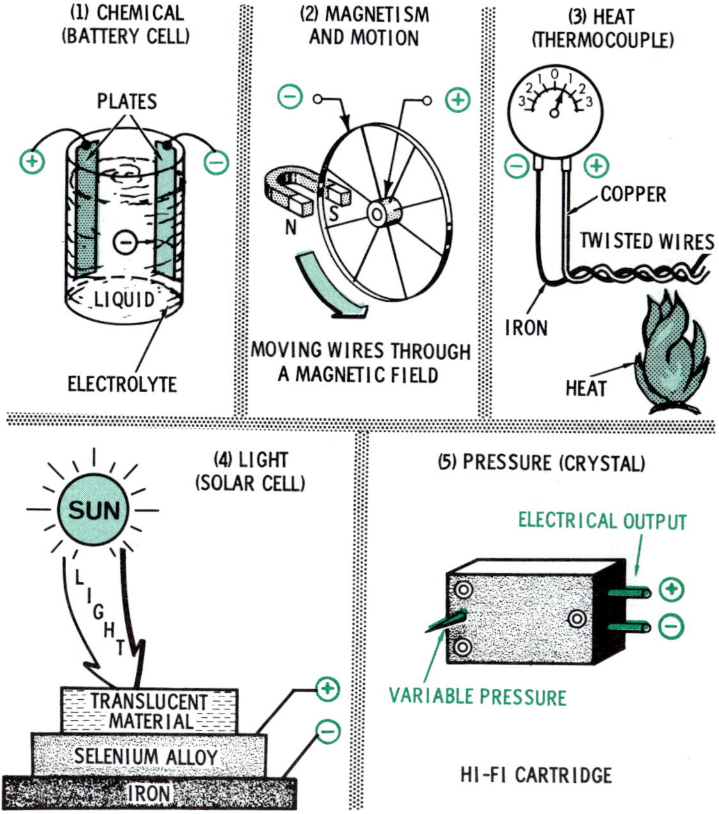


**Voltage sources.**

A battery or cell is a voltage source in which chemical action is used to develop a voltage potential. The cell is the basic unit. A battery consists of two or more cells placed together, end to end. For practical, uniform application, a dry cell is considered to be a 1.5-volt source. A lead-acid cell is a 2-volt source.

**Q49. What is the method of producing electricity not shown above? What is it called?**

**Your Answer Should Be:**

**A49.** The other method is **friction.** It is called **electrostatic electricity.**

### The Dry Cell

The most common form of a cell is the flashlight cell. The most common form of a battery is the automobile battery.

The cell is made with an outside protective covering which is an electrical insulator and, at the same time, is a mechanical support for all the other elements. Within this outer shell there are two electrodes separated and surrounded by an **electrolyte.** An electrolyte is any substance which, in solution, breaks down chemically into ions, thus allowing electric current to flow through it.

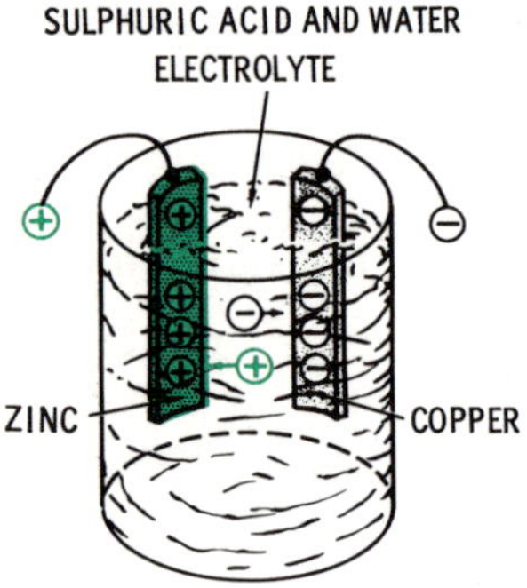


**A simple cell.**

The electrodes are generally made of two unlike metals. Ionization of the electrolyte causes current to pass between the two plates, or electrodes, in such a manner as to charge one positively and the other negatively.

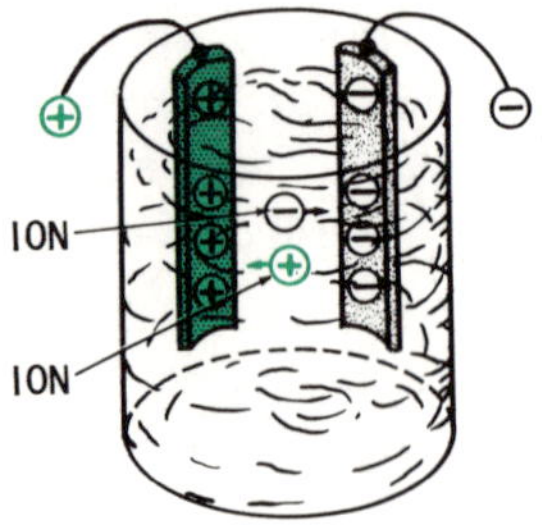


**Current flow in the cell.**

The dry cell is the most common type of DC-voltage source. It is constructed with an outside shell of zinc, a center rod of carbon, and a paste form of electrolyte which is a solution of ammonium chloride.

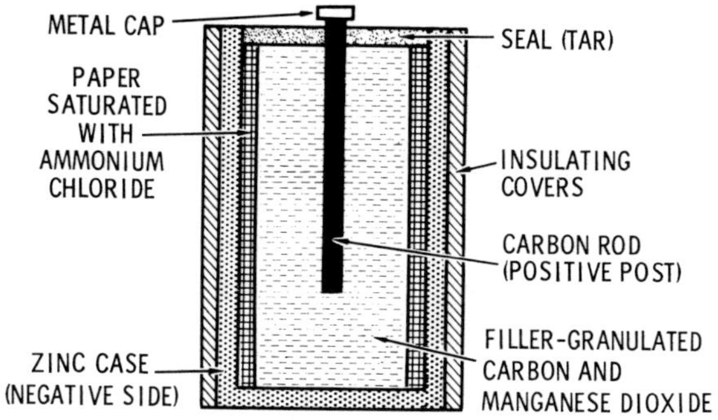


**Cutaway view of a dry cell.**

The zinc and the electrolyte are gradually "used up" as the dry cell supplies electricity. A dry cell is called a **primary cell** for this reason. The consumption of material in the cell is a result of chemical action between the zinc and the electrolyte. When the zinc case and the electrolyte have been used for some time, the chemical action which created the ion movement will decrease to a point where the cell is considered run-down (discharged).

**Q50. How does a cell produce voltage and current?**

**Q51. How would a dry cell be affected if the center rod were made of zinc and the outer shell of carbon?**

**Q52. How would a battery of 9 volts be constructed using dry cells?**

**Q53. What advantage is there to connecting the terminals of two cells in parallel (positive to positive, negative to negative)?**

**Q54. Why is a simple cell considered a primary cell?**

**Q55. How many lead-acid cells are in a 12-volt car battery?**

**Q56. If it takes the same amount of power ($I \times E$) for the starter to turn the engine in an automobile, in which system (6-volt or 12-volt) would more current have to be furnished to the starter motor?**

**Your Answers Should Be:**

**A50.** By **chemical action.**

**A51.** The cell would change the physical polarity of the posts.

**A52.** A 9-volt battery consists of **six 1.5-volt dry cells in series.**

**A53.** Two cells in parallel provide the same voltage but **supply twice as much current as a single cell.**

**A54.** A simple cell is a primary cell because **it is used up over a period of time** or arrives at a condition where there will be no more chemical action between the zinc and the electrolyte.

**A55.** A 12-volt car battery contains **six lead-acid cells.**

**A56.** **A 6-volt system** would have to furnish more current.

When a dry cell is discharged, it must be replaced. There are some discharge conditions in which a cell can be recharged, but these are very unreliable.

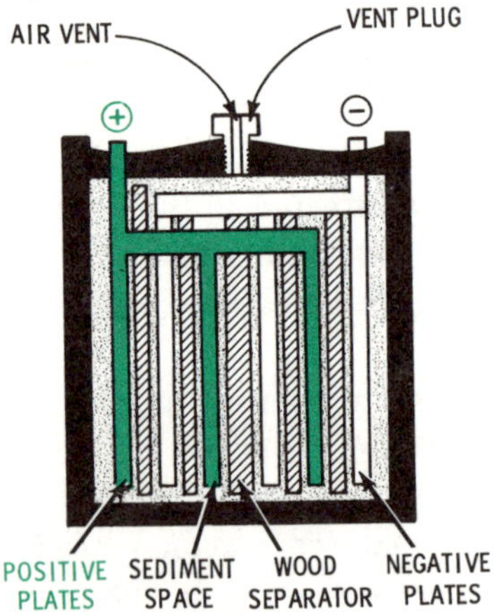


**A lead-acid cell.**

### The Lead-Acid Cell

The lead-acid cell is the type used in an automobile battery and has a normal output of approximately 2 volts. Since it can be recharged, a lead-acid cell is called a **secondary cell.**

A lead-acid battery must be constructed of a material that will not be eaten away by the highly corrosive electrolyte. For this reason, rubber or glass is normally used for the case that contains the cells and solution.

**Maintenance Precautions**

The maintenance of lead-acid cells requires the observance of certain safety and maintenance precautions to prevent injury and damage. Reasons for such care include: (1) the presence of dangerous sulphuric-acid electrolyte that will eat through clothing and skin, (2) generation of explosive fumes while battery is charging, (3) build-up of corrosion on connections and metal battery brackets.

The level of the electrolyte must be checked regularly to insure maximum effectiveness of the cell. This is a very

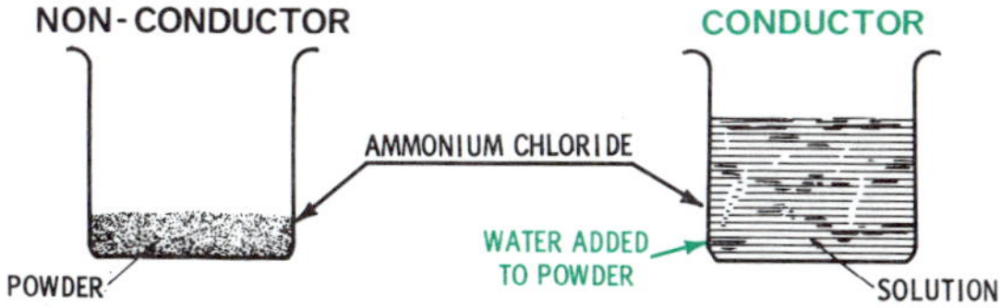


**An electrolyte.**

important monthly check for the automobile battery. If the electrolyte is permitted to get too low, the battery may become permanently damaged. The following is a list of items to be used when taking care of this type of battery:

1. A recording ledger listing each cell, the charge condition and date, electrolyte added, and the visual condition. Used as a maintenance history and to indicate changes in cell condition.
2. A cleaning brush, cleaning cloth, baking soda, and water. To remove corrosion on terminals and case.
3. White petroleum jelly to cover the terminals after they are cleaned and reconnected. Used to retard corrosion.
4. Charging equipment consisting of a battery charger, charge-rate meter, hydrometer, and an area separated from normally occupied areas so that a No Smoking rule may be observed.

**Q57. What is the difference between a cell and a battery?**

**Your Answer Should Be:**

**A57. A battery is a combination of cells** connected in series to form a voltage source.

The care of the lead-acid cell when it is in use or when it is being recharged requires special precautions.

1. Clean the battery or cell to remove corrosion. Coat the terminals with petroleum jelly to insure continued good connections.
2. Be careful when handling the battery to prevent spilling the electrolyte and to prevent internal damage to the plates. Remember that acid will cause serious burns and will develop a highly explosive atmosphere in a closed room.

A storage battery must be checked occasionally to see if it is charged properly, has sufficient electrolyte, is free of corrosion, and has clean tight connections. A **hydrometer** is a device commonly used to check the condition (specific gravity) of the electrolyte. In this way the condition of charge can be determined.

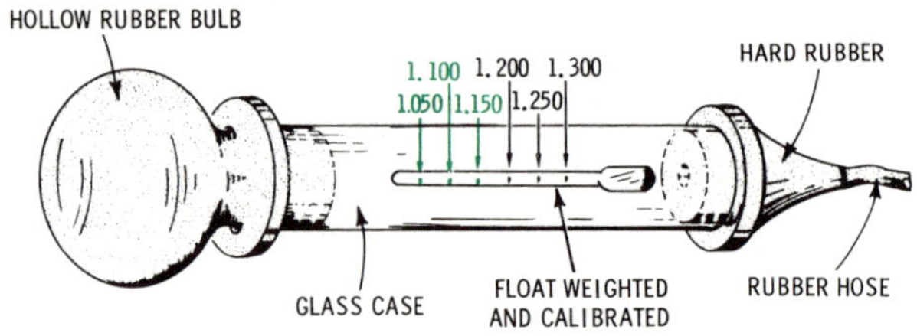


**A hydrometer.**

If the hydrometer reading is 1.250 to 1.300, the cell has a full charge. If the reading is 1.150 or less, the battery needs to be charged. If the reading is as low as 1.100, the battery is completely dead. A battery will rarely have the same reading for all cells. When using the hydrometer to check the battery, always keep the small tube or tip over the cell being checked so the electrolyte will drip back into the cell. After using the hydrometer, rinse it with water to remove any acid.

A dry cell should be purchased with a degree of caution. Even in storage, a dry cell will run down because of the chemical action of its components.

A quick method for checking an automobile battery is to try the horn. If the battery is weak, the sound of the horn will be weak. If a battery is being run down, recharged, run down, recharged again and again, its electrolyte level must be checked often. If a starter motor seems overly sluggish on cold mornings and the battery is known to be charged and in good condition, then all connections and cables should be checked. Many cars have been towed into a filling station or garage only to have a mechanic tighten a loose battery cable. When removing an auto battery, always remove the ground cable first. This will permit the use of metal tools. If the ground cable is removed first, the battery will be disconnected from the electrical circuit. Second, remove the "hot" side. This side will have no way of discharging to ground through a metal wrench if the ground post has had all connections removed first. When replacing the battery, attach the hot side first, then the ground cable. Making the ground connection last prevents arcing to the metal chassis of the auto during connection of the hot side.

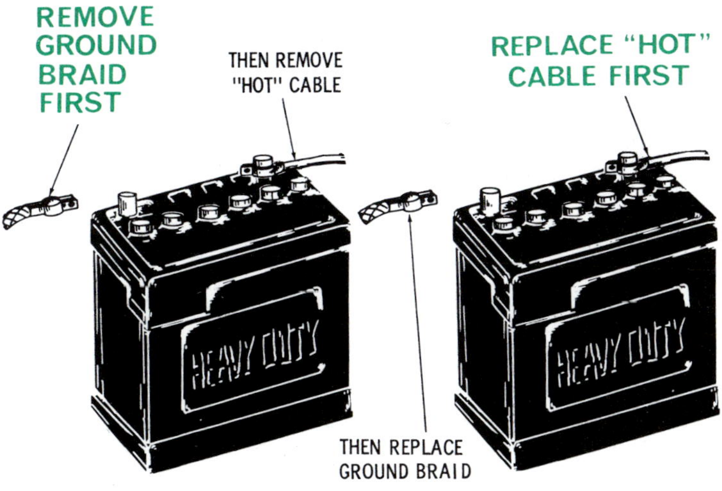


**Removing and installing a car battery.**

**Q58. What action takes place that causes a dry cell to run down on the shelf?**

**Your Answer Should Be:**

A58. **Chemical action** takes place within the cell even without a load.

## MAGNETIC VOLTAGE SOURCES

Electricity can also be generated by moving wires through a magnetic field. Also, electricity may be produced by moving the magnetic field through a number of wires.

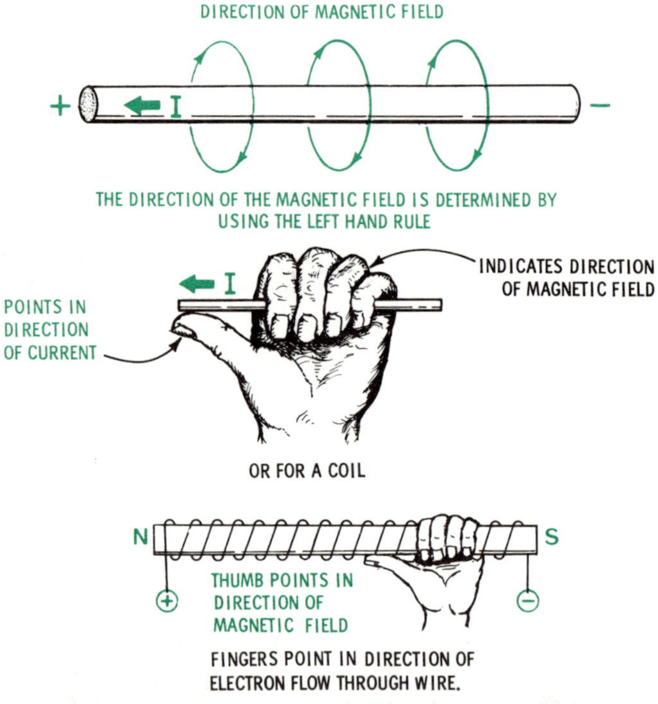


**Magnetism and current.**

To determine which way current will flow as a result of motion between a magnetic field and a wire, apply the **left-hand rule.** This rule will determine direction of current flow according to the electron theory. If the conventional current-flow theory is followed, use your right hand instead. Lines of force leave the north magnetic pole and enter the south pole.

The amount of current which can be generated by using magnetism and mechanical motion depends on several factors. One factor is the number of magnetic lines of force being cut. The greater this number, the greater the current

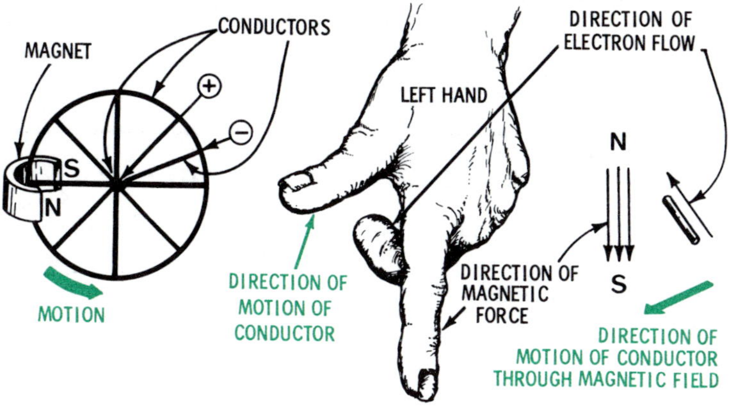


**Direction of current flow.**

will be. Another factor is the speed at which the lines of force are cut by the wires. The faster the relative motion between the two, the greater the force generating electron flow will be. Finally, the greater the number of wires cutting through the magnetic field, the greater will be the number of electrons that flow.

**Q59. In which direction will the electrons flow in the following diagram?**

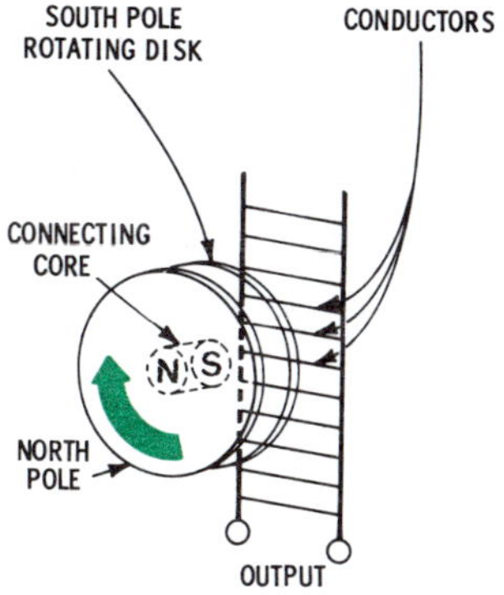


**Q60. Will an automobile generator produce more electricity when the car is traveling at 60 mph or at 20 mph?**

**Your Answers Should Be:**

**A59. Left to right.** (The effective direction of motion of the conductor is opposite to the direction of motion of the magnetic field.)

**A60.** The generator produces more electricity at **60 mph** than at 20 mph.

## HEAT-GENERATED VOLTAGES

Heat can be used to produce electricity by joining two different metals, heating the junction, and taking the output at the cooler end. Such a device is called a **thermocouple generator.**

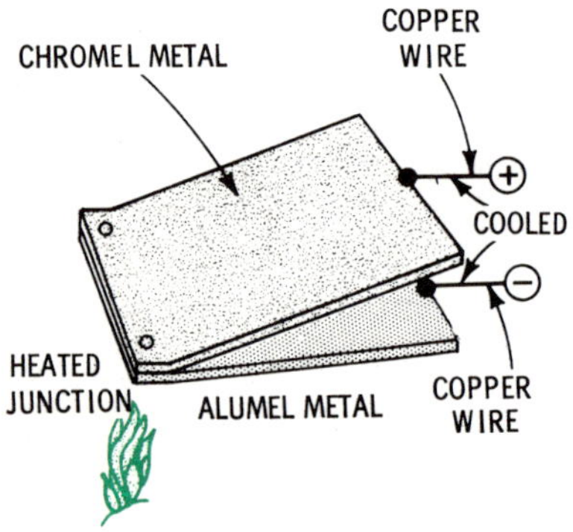


**A thermocouple.**

The thermocouple has many low-power applications. One such example is the radio-frequency current meter in the antenna circuit of a transmitter.

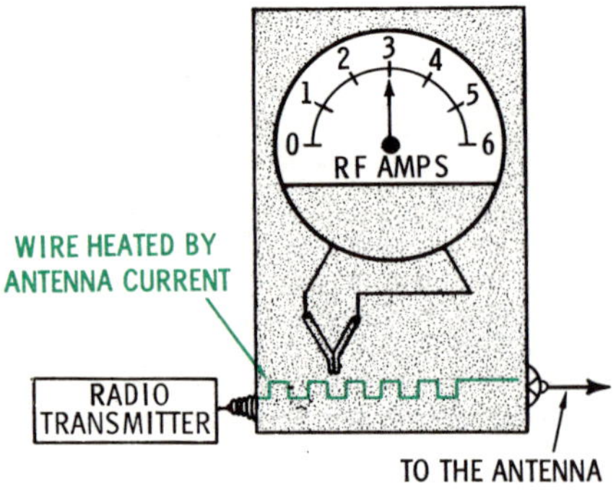


**An RF meter.**

A thermocouple produces an amount of voltage which is determined by the temperature difference between the two ends. The higher the difference, the greater the voltage.

*Transistor radio.*

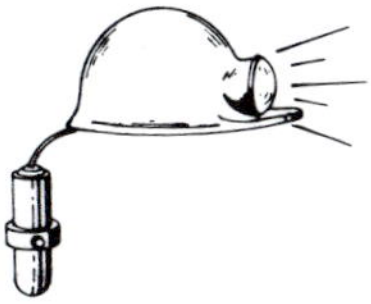

*Coal miner's lamp.*

**Thermocouple applications.**

Other applications for a thermocouple are in low-power devices such as small transistor radios and carbide lamps used by miners and hunters. A mixture of carbide and water produces a gas that will burn. The flame from a carbide burner heats the junctions of several thermocouples. The cooler ends are connected in series and produce a voltage which is then used to light an electric lamp.

**A thermocouple generator.**

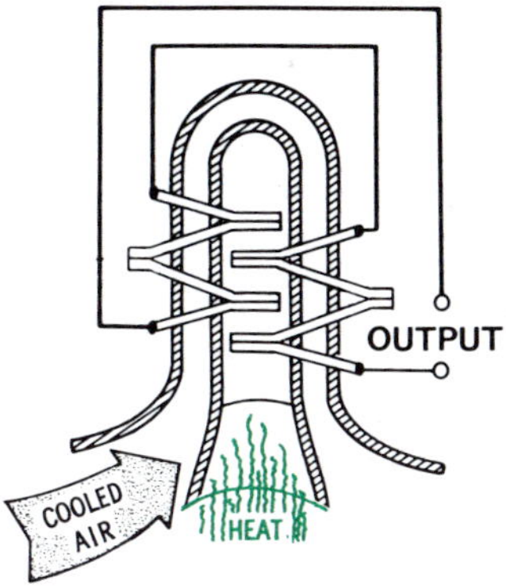


How can you make a thermocouple? One way is to twist the ends of an iron and a copper wire together. Heat the twisted junction over a flame. If the free ends are connected to a milliameter, a small current reading may be observed.

**Q61. Recalling the effect heat had on freeing the electrons in atoms, will the selection of thermocouple metals determine how much current will flow?**

**Q62. If electrons flow from one type of metal to the other in a thermocouple, is the current DC or AC?**

> **Your Answers Should Be:**
>
> **A61. Yes.** **A62. DC.**

## LIGHT-GENERATED VOLTAGES

A **solar cell** uses light to produce electricity. This device produces only very small amounts of voltage and current. Light striking certain materials causes electrons to move and develops a voltage at the terminals of the material. A **photoelectric cell** develops a voltage in a similar manner.

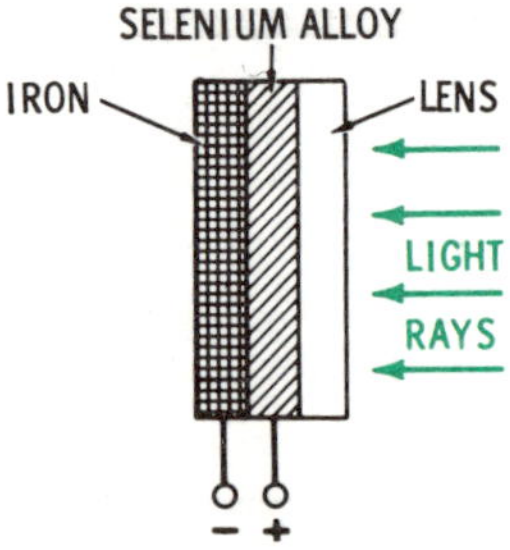


*Solar cell.*

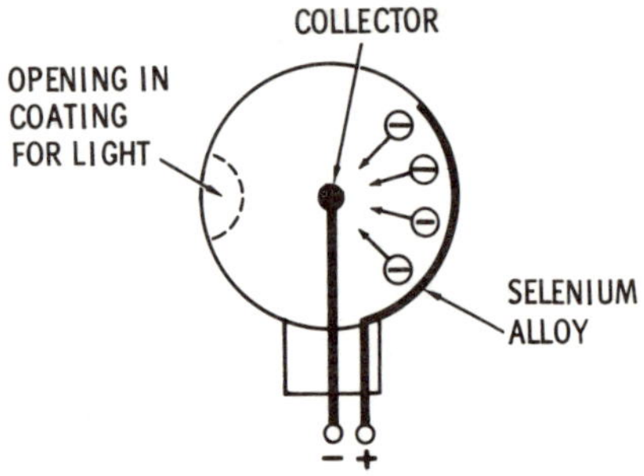


*Photoelectric cell.*

**Producing electricity with light.**

The photoelectric cell is often called a **PE cell.** A PE cell has many uses where a light beam can be broken, such as the automatic foul-line indicator on a bowling alley. Another application is an automatic lamp control. Here the PE cell is mounted so light from the sun will cause it to conduct and energize a relay. The energized relay turns the lamp off.

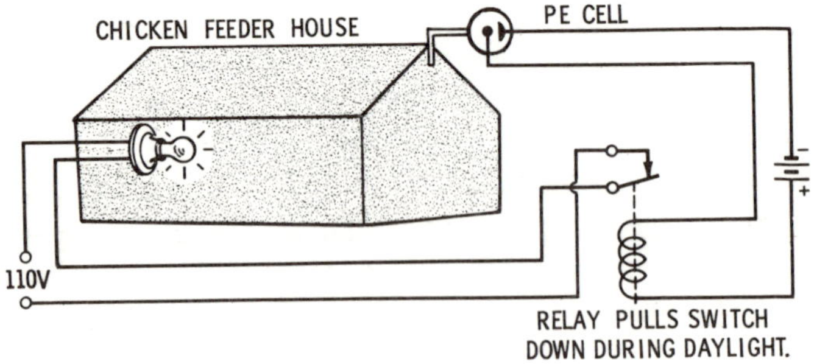


**Automatic lamp control.**

Selenium, used in solar and photoelectric cells, contains atoms that give up free electrons when struck by light. If iron is used as backing, the selenium will emit electrons into the iron. This will cause the iron to take on a negative charge, and the selenium to take on a positive charge. Wires connected to the two metals form the electrical source terminals.

In the solar cell, electrons travel from the selenium to the iron through their common bond. In the photoelectric cell, electrons move from the selenium through an evacuated area within the bulb to a center post.

Bonding, as used in a solar cell, is a common procedure employed in the electronics field. The bond is an electrical union similar to an electric weld. In fact, it is a bond created by a high current which causes the iron and the selenium to ionize and fuse together. When metals ionize they become very hot, just as in electric welding. Ionizing, as you recall, is the production of ions.

The vacuum bulb of the PE cell is the same type of evacuated bulb as that used for the electric light or the electron tube. The purpose of the vacuum is to allow electrons to travel from one location to another with a minimum of opposition from particles within the device. Air, for example, contains many millions of tiny particles which would hinder the flow of electrons.

The output of photoelectric and solar cells is very small. Therefore, the device which they control must require only a small amount of current. If they are to control larger objects, some form of electrical amplification must be used. The amount of voltage (or current) developed by these cells depends on the amount of light striking the selenium.

**Q63. Electricity is generated in the PE cell through the electron-releasing action of ________.**

**Q64. When _____ strikes selenium, selenium gives off __________.**

**Q65. A(an) ______ allows electrons to travel with a minimum of opposition.**

**Q66. The output of a PE cell is very (small, large).**

**Q67. The voltage output of a solar cell is _______ with more light and ____ with less light.**

**Your Answers Should Be:**

**A63.** Electricity is generated in the PE cell through the electron-releasing action of **selenium.**

**A64.** When **light** strikes selenium, selenium gives off **electrons.**

**A65.** A **vacuum** allows electrons to travel with a minimum of opposition.

**A66.** The output of a PE cell is very **small.**

**A67.** The voltage output of a solar cell is **greater** with more light and **less** with less light.

## PRESSURE-GENERATED VOLTAGES

**Crystals** of certain kinds produce electricity when pressure is applied to them. A crystal phonograph pickup is of this nature. It consists of two metal plates separated by a crystal, and a needle which is vibrated by the wavy variations in the groove of the record. These vibrations cause the crystal to be alternately squeezed and released, developing a small voltage across the terminals.

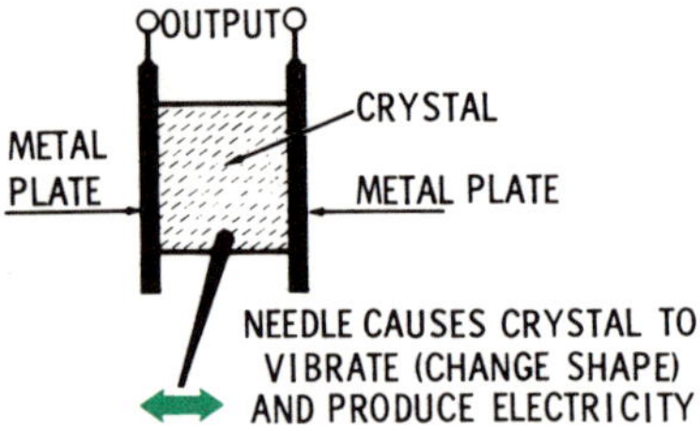


**A crystal phonograph pickup.**

The crystal most commonly employed for this form of electrical generator is Rochelle salt.

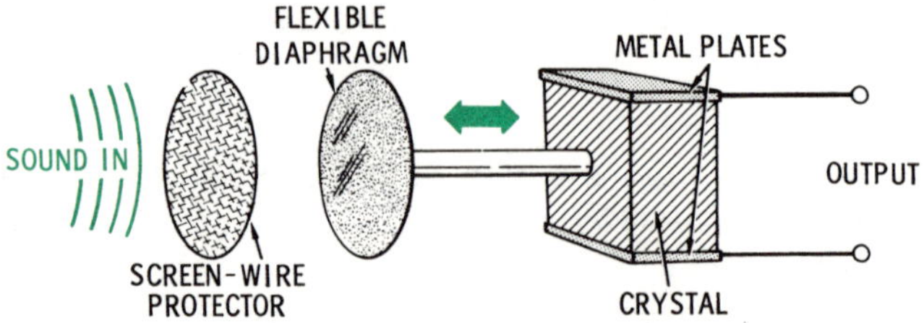


**A crystal microphone.**

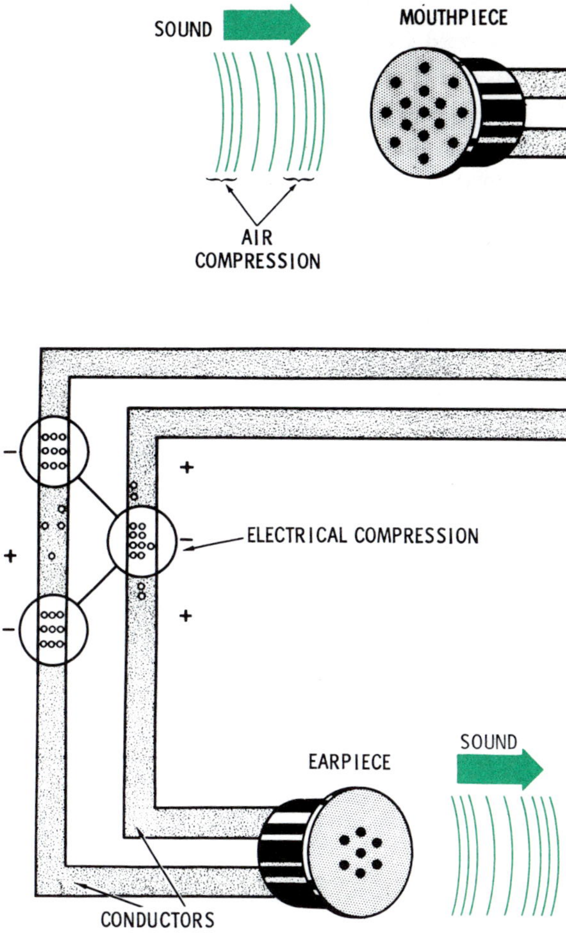


**Sound to electricity to sound.**

The ability of these crystals to produce a voltage when subjected to mechanical stress is called the **piezoelectric effect.** Conversely, when a voltage is applied to such crystals, a mechanical stress is produced.

**Q68. How does the amount of electricity produced by a crystal compare with the amount of pressure applied?**

**Q69. How does sound produce electricity in a sound-powered phone system?**

**Your Answers Should Be:**

**A68.** The **greater the pressure** is, the **greater the voltage.**

**A69.** The pickup element (mouthpiece) transfers the sound-generated vibrations to a crystal. Thus, electricity is produced. The resulting current is carried by conductors to the earpiece. The earpiece changes the electricity back into sound.

## RESISTANCE

The amount of current flow in a circuit is determined by how much voltage is applied and how difficult it is for current to flow. **Resistance** is the term used to describe the opposition offered to the flow of current.

How easily current will flow in a conductor depends on the number of free electrons available in a given area of the material. The longer the wire, the more resistance it has.

**Longer conductor—more resistance.**

At the same time, the larger the diameter, the less resistance a wire has.

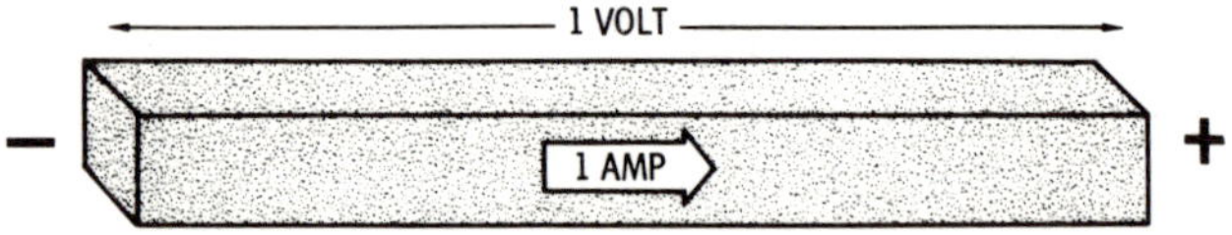


*1 unit of resistance.*

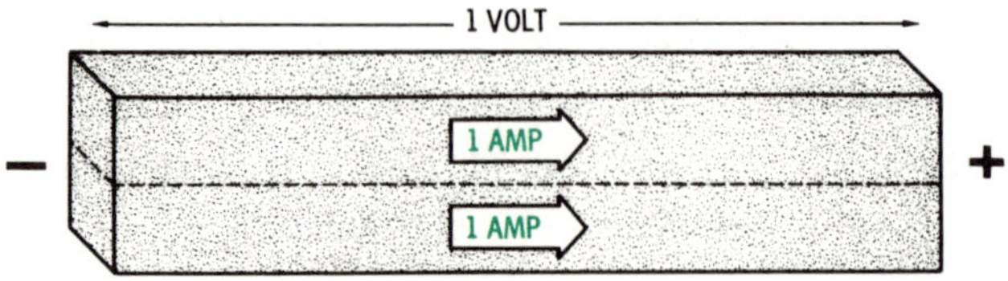


*½ unit of resistance.*

**More area—less resistance.**

### Computing Resistance of a Wire

The resistance of a wire can be determined as follows:

$$\text{Wire resistance is proportional to } \frac{\text{length}}{\text{area (cross-section)}}.$$

A wire 1,000 feet long will have twice the resistance of a 500-foot length having the same cross-sectional area. If both are the same length, a two-inch diameter wire will offer one-fourth as much resistance as a wire with a one-inch diameter.

Since the specific resistance of metals used as conductors can be determined, the preceding statement can be rewritten as a formula:

$$R = \rho \frac{L}{A}$$

where,

R is the resistance in ohms,
$\rho$ (Greek letter rho) is the specific resistance of the material per circular mil-foot,
L is the length in feet,
A is the cross-sectional area in circular mils.

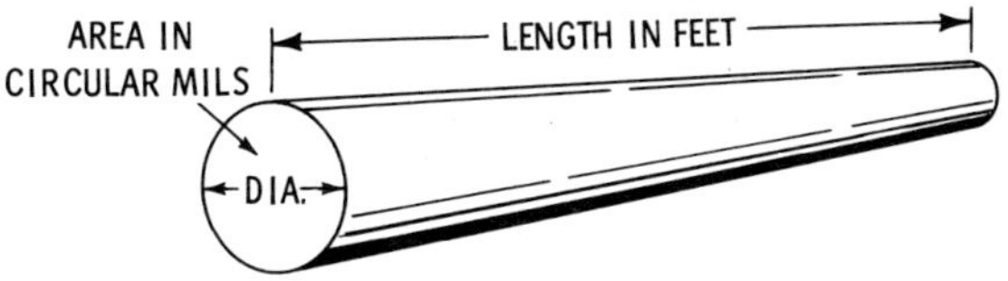


**Dimensions of a wire.**

A **circular mil** is used as a unit instead of square inches to measure cross-sectional area of a wire. Circular mils can be determined by squaring the diameter (in mils) of a wire. A mil is 0.001 of an inch.

If the above wire had a diameter of 0.06 inch, the area (A) would be $(60 \text{ mils})^2$, or 3,600 circular mils. If the diameter was 0.6 inch, A would equal $(600 \text{ mils})^2$, or 360,000 circular mils. If the two wires were of equal length, the second wire would be able to conduct 100 times more current, or offer 100 times less resistance, than the first wire.

**Specific Resistance**

The value of rho ($\rho$), the specific resistance of a conducting material, is expressed in ohms per circular mil-foot. The table below provides the specific resistance for several conducting materials. The ohmic values are given at 68°F. Values will be slightly higher at higher temperatures.

**SPECIFIC RESISTANCE ($\rho$)**

| Material | Ohms per circular mil-foot |
|---|---|
| Silver | 9.796 ohms |
| Copper | 10.370 ohms |
| Aluminum | 16.060 ohms |
| Tungsten | 33.220 ohms |
| Nichrome | 660.000 ohms |

Using this information, what is the resistance of 1,000 feet of copper wire having a diameter of 0.1 inch? This diameter is approximately the size of #10 electrical wire used for some applications in home wiring.

$$R = \rho \frac{L}{A}$$

$$\rho = 10.370 \text{ (from table)}$$

$$L = 1{,}000 \text{ feet}$$

$$A = 10{,}000 \text{ circular mils (100 mils squared)}$$

$$R = 10.370 \times \frac{1{,}000}{10{,}000} = 10.370 \times \frac{1}{10} = 1.037 \text{ ohms}$$

Unless the diameter of a wire is extremely small, its length very long, or its specific resistance high, the resistance of a conductor in a circuit is usually not considered. Therefore, unless stated otherwise, conductors will not be considered as part of the circuit resistance in problems given in this text.

However, care should be exercised in selecting the size of a wire in circuits where current may be high. For example, #12 (electrical gauge size) copper wire normally used in the home has a safe current-carrying capacity of only 20 amps. Higher current will cause it to develop too much heat.

## Valve Analogy

A valve used in the water system of a home is an example of resistance. When the valve is closed, water does not flow. If it is slightly open, a very small amount of water flows. The valve presents opposition (resistance) to the flow of the water. Even when the valve is completely open, less water will flow than if the valve were replaced with a pipe.

Will a large water pipe let more water flow than one of smaller diameter if the same amount of water pressure is applied? The answer is **yes.** By the same reasoning, a large wire will let more current flow than a small wire.

Will the length of the water pipe have any effect on the amount of water that flows out of the end? Yes. The longer the garden (lawn) hose you connect to the outside water valve, the less water pressure there will be at the end of the hose. Why is this? It is because the inside wall of the hose offers resistance to the flow of water.

As water travels through the hose, the water molecules rub against the side of the hose. The water molecules have motion but the hose does not, so the movement of the molecules is retarded. Slowing down the outer molecules also causes adjacent water molecules to decrease their speed. If the hose is of greater diameter, the center system of water molecules is less affected by the stationary outer wall.

Size of a wire has a similar effect on electron flow. A conductor of large diameter provides an easier (roomier) path for current than a conductor with a smaller diameter.

## Factors Affecting Resistance

The length and the cross-sectional area of a conductor determine how much resistance is in the circuit. Other things, such as the kind of material, temperature, and kind of electricity (AC or DC), also determine the resistance.

**Q70. Given a wire with a cross-sectional area of 2 circular mils and a length of 4 feet, and another wire with a cross-sectional area of 1 circular mil and a length of 2 feet, which wire will have the lesser opposition to current flow?**

**Q71. What is the difference between the resistance of a conductor and the resistance of an insulator?**

**Your Answers Should Be:**

**A70.** They will have the **same** opposition to current.

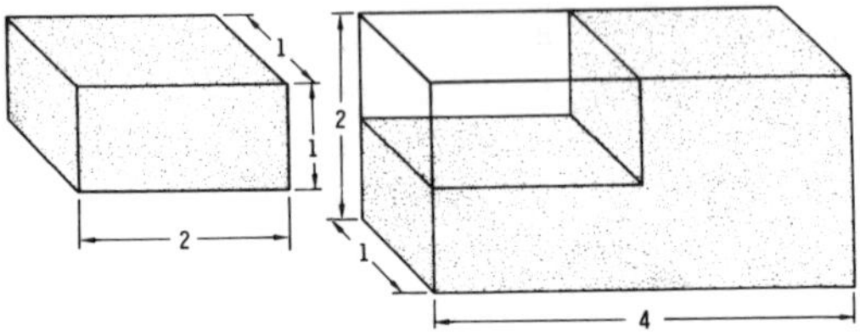


**A71.** The **conductor has a very low resistance** to current flow, and the **insulator has a very high resistance** to current flow.

## Resistance Units and Symbols

The standard unit of measurement for resistance is the **ohm.** One ohm of resistance in a circuit limits current to 1 ampere when 1 volt of electrical pressure is applied.

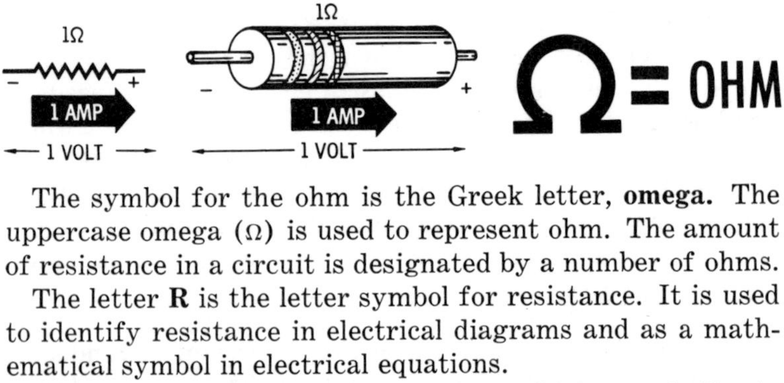


The symbol for the ohm is the Greek letter, **omega.** The uppercase omega (Ω) is used to represent ohm. The amount of resistance in a circuit is designated by a number of ohms.

The letter **R** is the letter symbol for resistance. It is used to identify resistance in electrical diagrams and as a mathematical symbol in electrical equations.

Resistance units have several values which are similar to those for voltage and current.

| Resistance Designation | Meaning |
|---|---|
| 1 ohm (Ω) | unit of resistance |
| 1 kilohm | 1 KΩ = 1,000 Ω |
| 1 megohm | 1 MΩ = 1,000,000 Ω |
| 1 milliohm | 1 mΩ = 0.001 Ω |

## Resistors and Resistive Components

Resistance is not limited to conductors alone. Electronic components, called **resistors,** are manufactured to have a specific amount of resistance. Many resistors of several values are used in radios and TV sets. Also, there are many applications for resistors in commercial and military equipments.

What is the purpose of resistors in these circuits? They are used to limit current flow and to develop voltages of lesser potential than the source. Resistors are manufactured in many different forms and many different values. They may also be of a fixed or variable value.

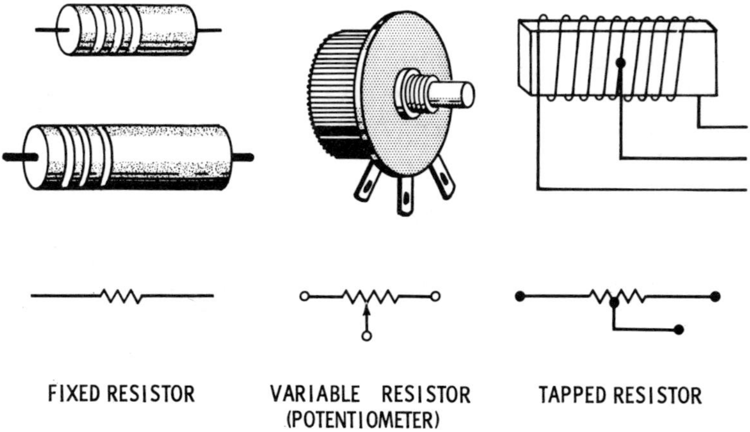


**Resistor types and symbols.**

**Q72. The unit of measurement for resistance is the ---.**

**Q73. The ____ is the symbol for resistance value.**

**Q74. What are resistors used for?**

**Q75. As an exercise, write the following resistance values in the shortest possible form.**

**(a) 37,500 Ω** **(b) 375,000 Ω**

**(c) 1,030,000 Ω** **(d) 103 Ω**

**(e) 99,000 Ω** **(f) 146,210 Ω**

**(g) 4,900 Ω** **(h) 57,200 Ω**

**Your Answers Should Be:**

A72. The unit of measurement for resistance is the **ohm.**

A73. The Ω is the symbol for resistance value.

A74. To **limit current flow** and to **develop voltages smaller than the source voltage.**

A75. (a) **37.5KΩ** (b) **375KΩ**

(c) **1.03MΩ** (d) **103Ω**

(e) **99KΩ** (f) **146.21KΩ**

(g) **4.9KΩ** (h) **57.2KΩ**

## Resistor Values

The material from which resistors are made is of such a nature as to produce the desired values. Most resistors are manufactured from wire or from a mixture of materials.

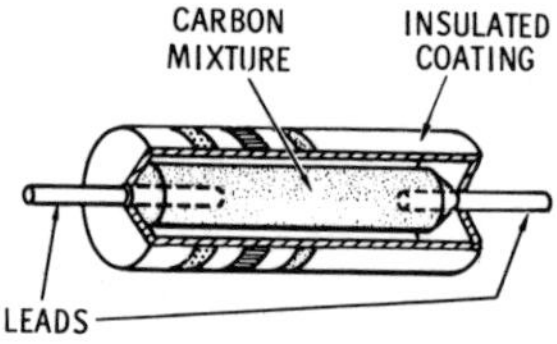


*Carbon-composition.*

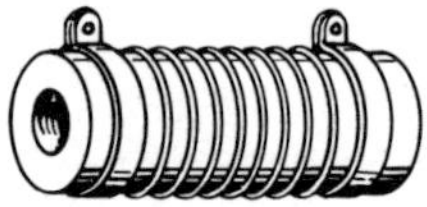

*Wirewound (tubular).*

**Carbon and wirewound resistors.**

## Reading and Measuring Resistors

Resistors are color coded to permit determination of their resistance value. You will find three or four colored bands around the resistor. The bands will be nearer one end than the other. The first band is nearest the end of the resistor.

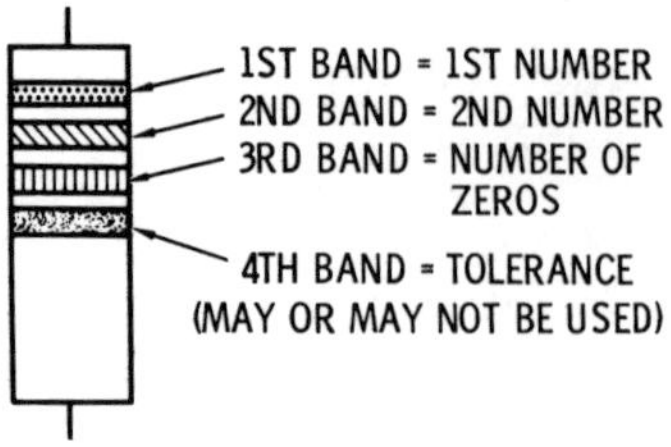


**Resistor color bands.**

Ten colors are used to represent each of the digits in the decimal system. These colors are black, brown, red, orange, yellow, green, blue, violet, gray, and white. They represent the digits 0 through 9, in that order.

The first two colors on the resistor (the two nearest the end) represent the first two numbers of the resistance value. The third color band represents the number of zeros that follow the first two numbers. The fourth band (if there is one) determines how much the actual resistance can vary from the indicated value of the resistor. The difference between the indicated value (by color bands) and the actual value is called the resistor **tolerance.**

The fourth band will be either gold or silver. If it is gold, the tolerance of the resistor is plus or minus 5%; if it is silver, the tolerance is plus or minus 10%. The absence of a fourth color band indicates the resistor tolerance is plus or minus 20%. For example, the actual value of a 68,000-ohm resistor having only three color bands is between 54,400 ohms and 81,600 ohms. (20% of 68,000 ohms is 13,600. The tolerance limits are therefore 68,000 + 13,600 = 81,600 ohms, and 68,000 — 13,600 = 54,400 ohms.)

**Q76. What are the digits for the following colors? (a) Blue. (b) Gray. (c) Orange. (d) Black.**

**Q77. What are the values of the following resistors?**

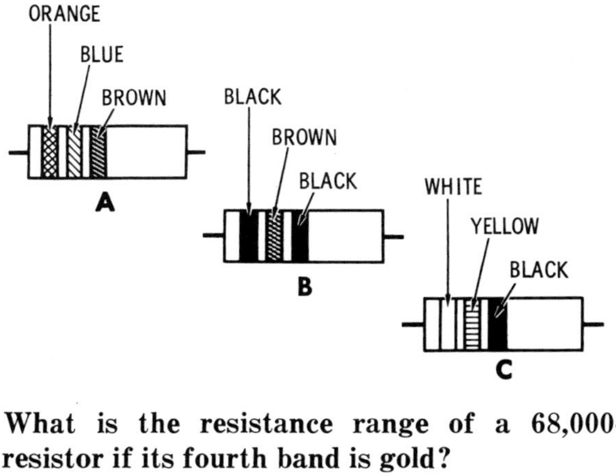


**Q78. What is the resistance range of a 68,000-ohm resistor if its fourth band is gold?**

**Q79. What is the resistance range of a 68,000-ohm resistor if its fourth band is silver?**

**Your Answers Should Be:**

**A76.** (a) Blue is **6.** (b) Gray is 8. (c) Orange is **3.** (d) Black is **0.**

**A77.** (a) **360 ohms.** (b) **1 ohm.** (c) **94 ohms.**

**A78.** **64,600 to 71,400 ohms** (68,000 plus or minus 3,400 ohms).

**A79.** The resistor would read somewhere **between 61,-200 ohms and 74,800 ohms.**

## WHAT YOU HAVE LEARNED

1. All matter is made up of molecules.
2. A molecule is the smallest possible particle that still retains the same physical and chemical characteristics of a substance.
3. Molecules are made up of atoms.
4. An atom is the smallest portion of an element which exhibits all the properties of that element.
5. An atom is made up of even smaller particles called protons, electrons, and neutrons.
6. Electrons have a negative charge. They revolve around the center cluster in orbit zones.
7. The center cluster of an atom is called the nucleus and consists of protons and neutrons bonded together.
8. A proton has a positive charge.
9. A neutron has a positive and negative charge to make it neutral.
10. Different elements have different numbers of electrons and protons.
11. The electrons orbit around the nucleus in zones which may or may not be filled.
12. The first two orbit zones are filled when they have 2 and 8 electrons, respectively. Subsequent zones are filled with 8, 18, or 32 electrons.
13. If an atom has a different number of electrons than

protons, it has an electrical charge (either positive or negative).

14. A charged atom is called an ion. It is a negative ion if it has an excess of electrons. It is a positive ion if it has a deficiency of electrons.
15. An element whose atom does not have completely filled outer-orbit zones is called a free-electron type element.
16. A free-electron type element requires less energy from a voltage source to cause transfer of electrons from one atom to another than does an element whose orbit zones are filled.
17. To have electrical movement, energy must be applied to an atom.
18. To be of value, the energy must be applied so as to direct electrical movement.
19. Voltage (electrical pressure) is the most common form of control.
20. Voltage is a difference of electrical charge between two points.
21. The difference in electrical charge between two points establishes a force.
22. The amount of force is determined by the amount of charge at each point and the distance between them.
23. The production of electricity is the result of controlled movement of ions or electrons. This movement is called current.
24. Electricity, once produced, can be transferred from one location to another.
25. Electrically, materials may be classified as conductors or insulators.
26. Conductors can serve as a path for the transfer of electricity.
27. Insulators present a very difficult path for the transfer of electricity.
28. The human body is a conductor of electricity.

    **NOTE:** For this reason, any person near electricity should take every precaution possible in order not to become a part of the electric circuit!

29. Electricity may be produced by friction.
30. When electricity is produced as a result of one object rubbing against another, it is called static electricity.
31. Lightning is an example of static electricity.
32. The production of electricity is the result of a collection of more positive charges at one point than at another, or a collection of more negative charges at one point than at another.
33. An accumulation of an electrical charge (voltage) can be measured with a voltmeter.
34. The amount of current flowing through a circuit when voltage is applied can be measured with an ammeter.
35. The amount of force between two charged bodies can be calculated by using the expression:

$$F = \frac{Q_1 \times Q_2}{d^2}$$

36. Graphical representation of electric force and electrostatic fields is used to indicate the direction an electron will move when acted on by charged bodies.
37. Electric current is the flow of either ions or electrons.
38. There are two major current theories employed in the electrical and electronic fields.
39. One is electron-theory current flow and the other is conventional current flow.
40. Electron-theory current flows from the negative terminal of a voltage source through the circuit, and back to the positive terminal of the source.
41. Conventional current (ions) flows from the positive terminal, through the circuit, and back to the negative terminal of the source.
42. The standard unit for current is the ampere.
43. "a" is used to indicate ampere.
44. "I" is used to indicate the current flowing in a circuit.
45. Voltage is electrical pressure.
46. Voltage may be called potential difference, electromotive force (emf), or voltage potential.
47. The standard unit for voltage is the volt.

48. "V" is used to indicate volt.
49. "E" is used to represent voltage in a circuit.
50. Electricity can be produced by means of chemical action, heat, light, magnetism, pressure, or friction.
51. A battery is a device which uses chemical action to produce electricity.
52. A thermocouple employs heat to produce electricity.
53. An automobile generator uses magnetism to produce electricity.
54. A crystal pick-up of a record player converts pressure into electricity.
55. The amount of current flow in a circuit is determined by how much voltage is applied and the amount of resistance in the circuit.
56. The standard unit of measurement for resistance is the ohm.
57. The Greek letter omega ($\Omega$) is used to represent ohm.
58. "R" is used to represent resistance in a circuit.
59. Colored bands around a resistor indicate the value of resistance in ohms and the tolerance.

# 2

# The Simple Electrical Circuit

## *What You Will Learn*

**You are now going to learn about basic circuits. You will be shown the application of Ohm's law and the electrical power relationships which exist between voltage, current, and resistance. From this information you will be able to identify basic circuits; visualize and describe voltage, current, and resistance relationships; determine power relationships; and construct a basic circuit.**

### BASIC CIRCUITS

A circuit consists of a closed path through which an electric current flows. It is the route followed by current as it

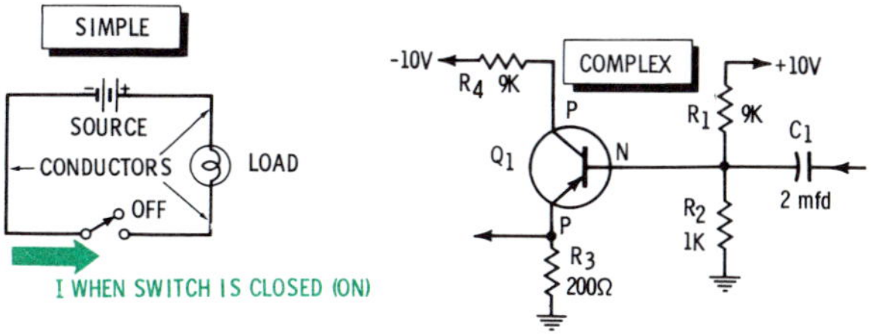


**Circuit examples.**

travels from the voltage source through the conductors and the load and returns to the source. This includes the path from terminal to terminal through the source.

## Connecting a Source to a Load

A circuit must form a **complete** (unbroken) loop for current to flow. Current flows through a circuit in somewhat the same manner as a bicycle chain makes a complete loop between the driving gear (source) and the wheel (load).

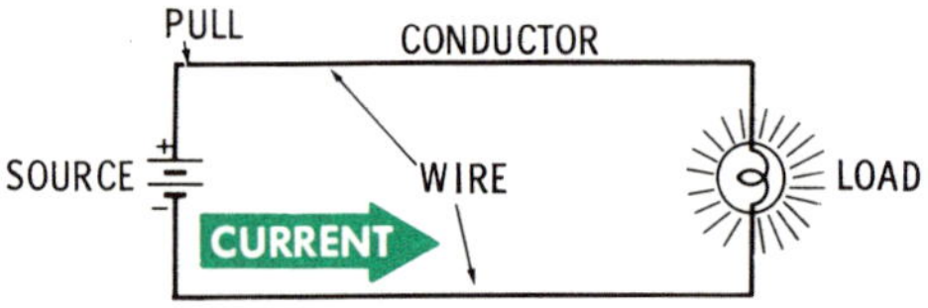


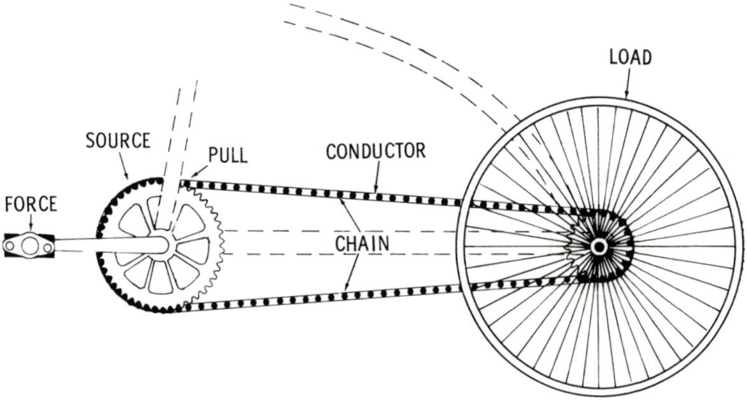


An electrical/electronic circuit must form a complete loop from the voltage source to the load and back to the source.

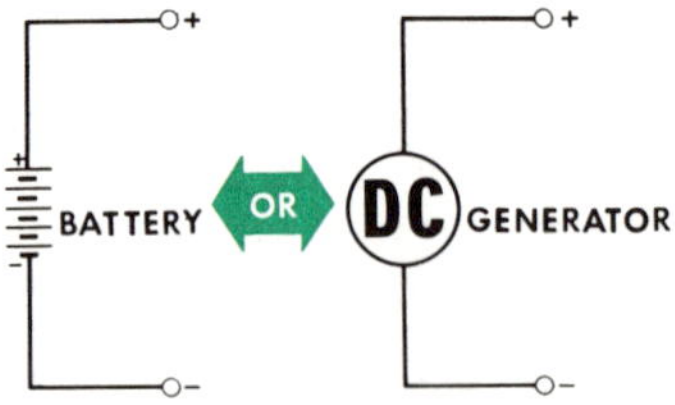


**DC voltage sources.**

Symbols used for DC voltage sources in schematic diagrams are normally those of a battery or a DC generator.

The current-carrying paths between the terminals of a load and its voltage source are called **conductors.** A conductor can be a wire (usually copper), or the heavy metal that forms part of the equipment or device can serve as one of the paths between the load and voltage source.

## SWITCHES

Switches have been developed so that a circuit may be opened and closed with little effort or danger to the operator. Some switches are normally closed. When these switches are actuated, they **open,** or break, the circuit. Other types of switches are normally open. When actuated, they **close** (make) a circuit. Such switches are maintained in their normal positions by springs.

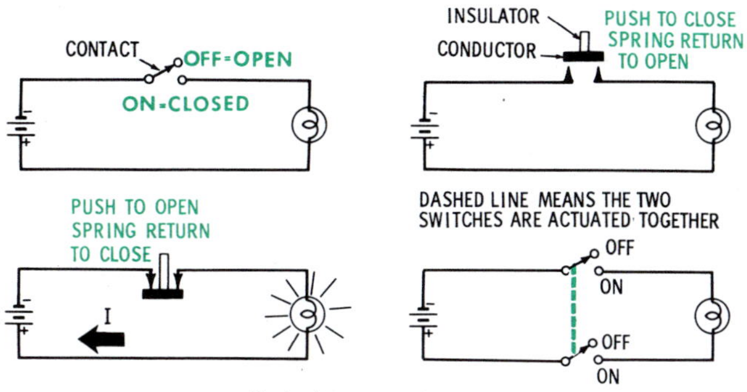


**Switch operation.**

**Q1. If a lamp is mounted on an insulator in an automobile, how many wires must be connected to the lamp socket?**

**Q2. There are no wires in a flashlight. How is the circuit completed?**

**Q3. A tail lamp on an automobile is mounted on a very rusty body. If the lamp lights only part of the time, what is the possible cause of the trouble?**

**Q4. A completed circuit containing a voltage source will always have -------- -------- through it.**

**Q5. A circuit is open if current (does, does not) flow.**

**Q6. A lamp circuit is ------- if the lamp lights.**

**Q7. --------- are used to open and close circuits.**

**Your Answers Should Be:**

**A1.** There must be **two** wires connected to the lamp socket. One is connected to a "hot" terminal and the other to the chassis.

**A2.** The flashlight shell (case) forms one conductor from the lamp to the batteries. A battery terminal in direct contact with the lamp forms the other conductor.

**A3.** The probable cause is a **poor** or **open conductor** (the rusty body).

**A4.** A complete circuit containing a voltage source will always have **current flowing** through it.

**A5.** A circuit is open if current **does not** flow.

**A6.** A lamp circuit is **closed** if the lamp lights.

**A7.** **Switches** are used to open and close circuits.

The **knife switch** was probably the first switch used to any great extent. A basic requirement for the knife switch is the alignment of the two ends. This is required to permit a movable blade to make positive connection with stationary contacts. The switch is usually mounted on insulators or on an insulated board.

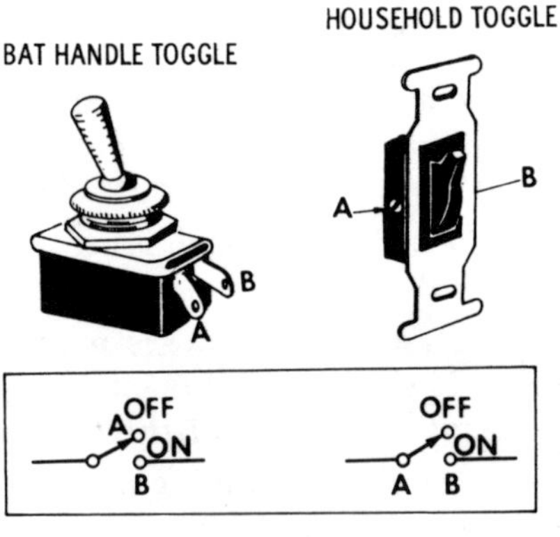


**The common toggle switch.**

A **toggle switch** is an improved knife switch. It uses a spring-loaded mechanism to open and close the contacts.

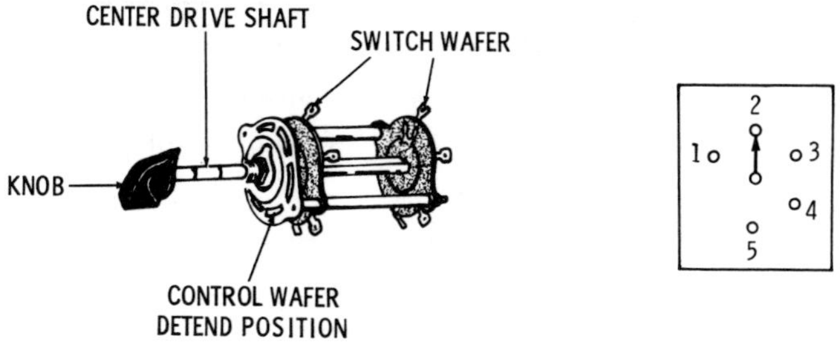


**The wafer switch.**

A **wafer switch** is a rather recent development resulting from the many advances in electronic and electromechanical devices. Electromechanical devices include equipment that converts electrical energy to mechanical energy or mechanical energy to electrical energy. One advantage of the wafer switch is its ability to be rotated from one position to another, providing a multiple switching action. Wafer switches are made to rotate either a full 360°, or part of a turn. Another advantage is that more than one wafer may be actuated with a single shaft.

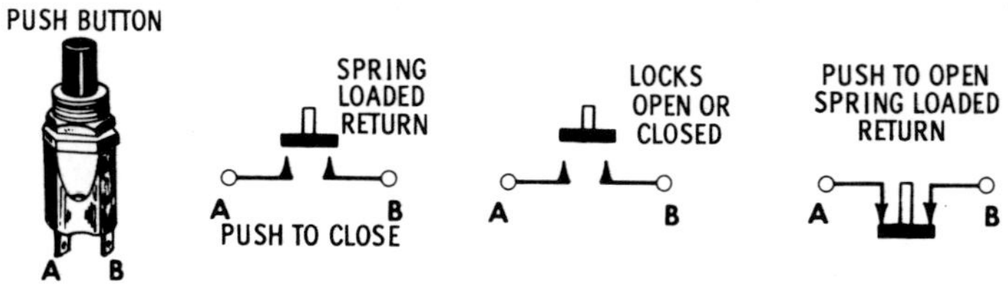


**The push-button switch.**

A **push-button switch** uses a button which must be pushed to form a closed (or open) condition. One type remains closed (or open) until the next push. Another kind is the **momentary** push button. It stays closed (or open) as long as the button is depressed.

Regardless of the style, a switch is used to fulfill only one purpose—to open or close a circuit.

**Q8. What is the purpose of a switch in an electrical circuit?**

**Q9. What unique feature does the wafer switch have over other switches discussed in this chapter?**

**Your Answers Should Be:**

A8. A switch is used to **open and close circuits.**

A9. A wafer switch is the only one that employs **rotary motion** for positioning.

## OHM'S LAW

Ohm's law expresses the precise relationship that exists between voltage, current, and resistance. Relationships concerning this law for DC circuits will be used for evaluation of all other electrical circuits. You should therefore study the next few pages carefully.

### Basic Relationships

Which circuit below will have the greater current flow?

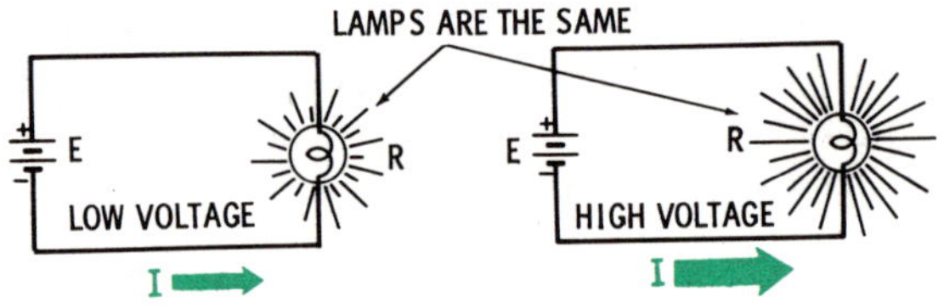


**Ohm's law.**

The circuit with the lower voltage will not have as much current flow as the other. The current will therefore be greater in the circuit on the right.

In which of the following circuits will current be greater? The two voltages are equal.

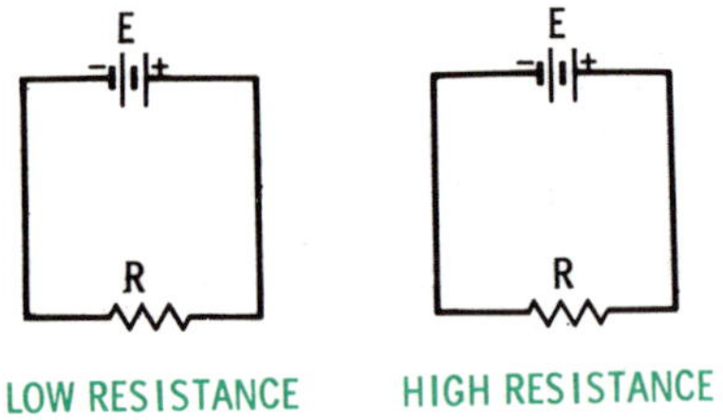


**Voltage constant—resistance varied.**

Current cannot flow as easily through a high resistance as it can through a low resistance. Thus, the circuit on the left will have the greater current flow. The circuit with the higher resistance will require a higher voltage source to cause an equal amount of current to flow.

## Explanation of Ohm's Law

Current will decrease if voltage is decreased or resistance is increased. The reverse is also true. Current will increase if voltage is raised or resistance is reduced. Therefore, Ohm's law can be stated as an equation, with current (I) being made equal to a ratio between voltage (E) and resistance (R).

$$I = \frac{E}{R}, \text{ or amperes} = \frac{\text{volts}}{\text{ohms}}$$

This statement (equation) is true for the current-voltage-resistance relationship in any DC circuit. If R (in ohms) remains the same, and E (in volts) is increased by a factor of two, five, ten, or any other number, then I (amperes) will be increased by the same factor. A decrease in the number of volts will result in a proportional decrease in amperes. In other words, to determine I, E is divided by R.

$$I = \frac{E \text{ (10 volts)}}{R \text{ (2 ohms)}} = \frac{10}{2} = 5 \text{ amps}$$

The same reasoning holds true for changes in resistance. For a given voltage, current will be twice as large if resistance is decreased to one-half its former value. If the resistance is increased four times, current will become one fourth as large.

**Q10-15. Find I in each of the six circuits below.**

**Q10.**

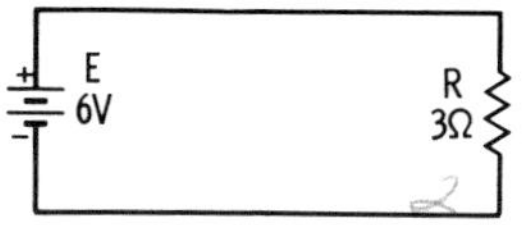


**Q11.**

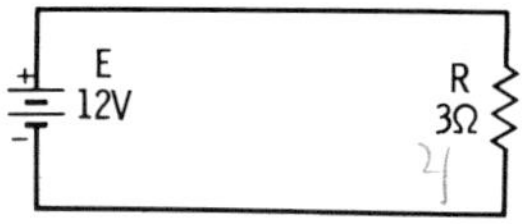


**Q12.**

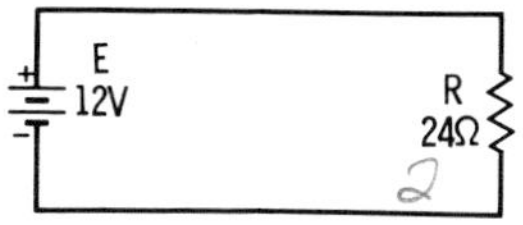


**Q13.**

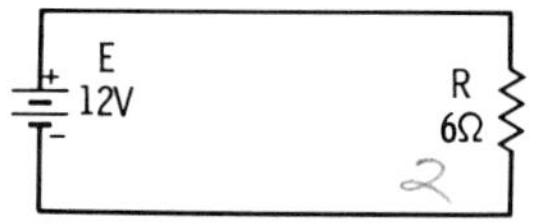


**Q14.**

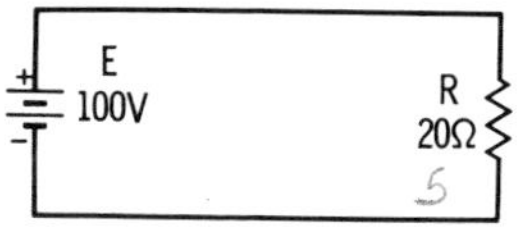


**Q15.**

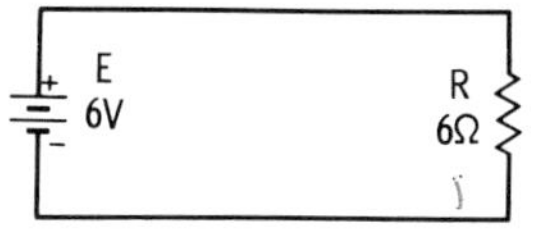

**Your Answers Should Be:**

**A10. 2 amps.**

$$I = \frac{E}{R} = \frac{6 \text{ volts}}{3 \text{ ohms}} = 2 \text{ amps}$$

**A11. 4 amps.** Note the amount of increase in current when the voltage was doubled.

**A12. 0.5 amp**

**A13. 2 amps.** Note how much current increased when the resistance was reduced by one fourth.

**A14. 5 amps**

**A15. 1 amp**

## Application of Ohm's Law

Ohm's law can be used to find either current, voltage, or resistance in a circuit if the other two factors are known.

**Solving for Current**—You have already seen how current in a circuit can be determined by dividing the circuit voltage by the load resistance. I is equal to E divided by R.

**Solving for Voltage**—In the chapters which follow, you will see the need for determining voltage when current and resistance are known. For example, how much voltage is required to force a current of 2 amperes through a load resistance of 50 ohms? The Ohm's-law equation would be:

$$I = \frac{E}{R}, \text{ or } 2 \text{ amps} = \frac{? \text{ volts}}{50 \text{ ohms}}; E = 100 \text{ volts}$$

Arithmetic reasoning tells you that voltage units must be twice the number of resistance units to permit 2 amperes to flow. Or, you multiply the number of ohms by the number of amperes to obtain the number of volts. E is equal to I multiplied by R. By using a mathematical rule, you multiply both sides of the basic equation by R to obtain the same equation for voltage.

$$(R) \times I = \frac{E \times (R)}{R}, \text{ or } IR = E$$

The R's on the right side of the equation cancel, leaving E equal to IR (I multiplied by R). If 0.5 amp is flowing through 20 ohms, E applied (IR) is 10 volts.

**Solving for Resistance**—If the voltage and current in a circuit are known, its resistance can be determined by similar reasoning. What is the resistance of a coil that permits 2 amperes of current to flow when connected to a 6-volt battery?

$$I = \frac{E}{R}, \text{ or } 2 \text{ amps} = \frac{6 \text{ volts}}{? \text{ ohms}}; R = 3 \text{ ohms}$$

To obtain 2 amperes, 6 volts must be divided by 3 ohms. In other words, resistance can be determined by dividing voltage by current. Mathematically, the equation for resistance is obtained in the following manner:

$$I = \frac{E}{R} \text{ (transpose I and R)} : R = \frac{E}{I}$$

**Q16-23. Solve for the quantity indicated in the circuits below.**

**Q16.**

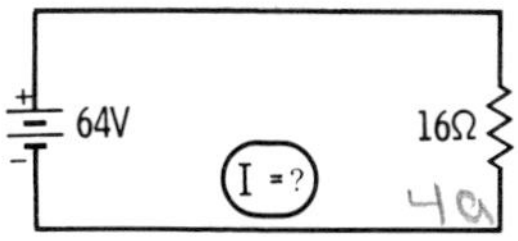


**Q17.**

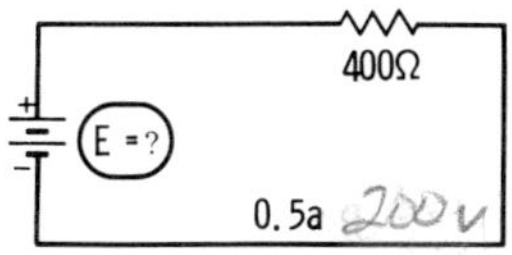


**Q18.**

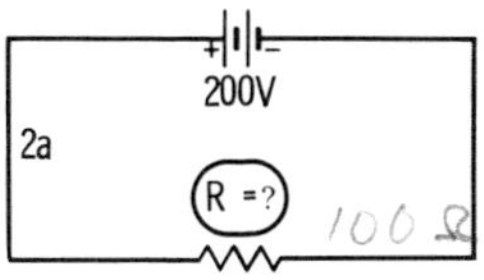


**Q19.**

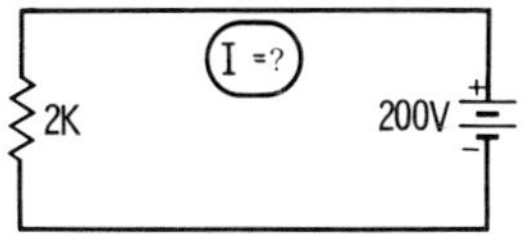


**Q20.**

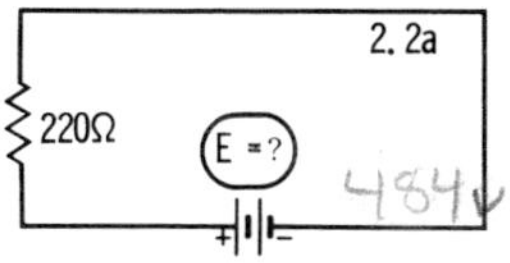


**Q21.**

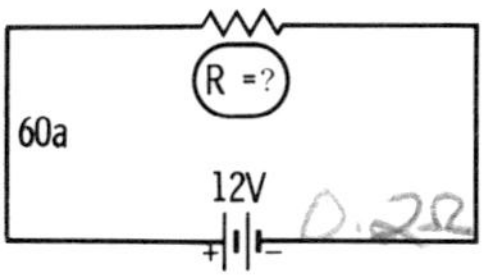


**Q22.**

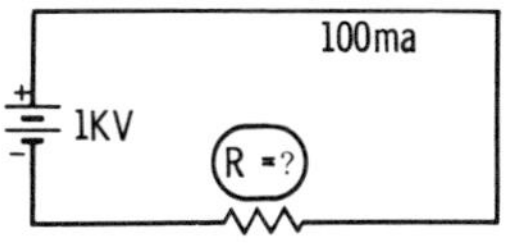


**Q23.**

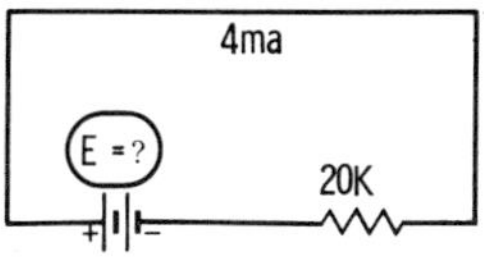

**Your Answers Should Be:**

A16. $I = \frac{E}{R} = \frac{64}{16} = $ **4 amps**

A17. $E = IR = (0.5)(400) = $ **200 volts**

A18. $R = \frac{E}{I} = \frac{200}{2} = $ **100 ohms**

A19. $I = \frac{E}{R} = \frac{200}{2{,}000} = $ **0.1 amp**

A20. $E = IR = (220)(2.2) = $ **484 volts**

A21. $R = \frac{E}{I} = \frac{12}{60} = \frac{1}{5}$, or **0.2 ohm**

A22. $R = \frac{E}{I} = \frac{1{,}000}{0.1} = 10{,}000$, or **10K ohms**

A23. $E = IR = (0.004)(20{,}000) = $ **80 volts**

## VOLTAGE DROP

**Voltage drop** is a term which can be misleading. The word "drop" may lead you to believe that an amount of voltage is lost. This is not true, however; voltage never disappears. Voltage drop merely refers to the manner in which the source voltage is distributed (dropped) throughout the circuit.

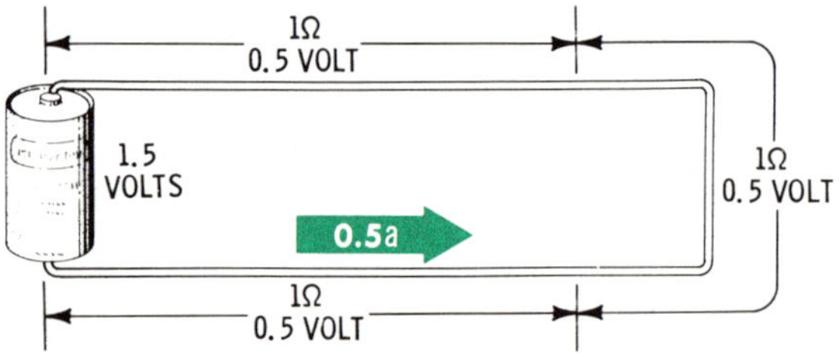


**Voltage drop in a circuit.**

All of the source voltage is distributed proportionally across the resistance in a circuit. The voltage drop between any two points in a circuit can be determined by the ratio of the individual resistance to the total resistance of the circuit. The diagram, for example, shows a 3-yard length of wire connected to the terminals of a 1.5-volt cell. The resistance of the wire is one ohm per yard, or a total of three ohms for the circuit. Current for the circuit (E/R) will be 0.5 amp.

The total 1.5 volts will be dropped across the total length of the wire. Since E = IR, the voltage distribution is often called the **IR drop.** This version of Ohm's law can be used to determine the voltage drop across any portion of a circuit.

Since one yard of wire presents one ohm of resistance to the current flowing through it, its IR drop (share of the circuit voltage) is 0.5 amp × 1 ohm, or 0.5 volt. A yard of wire represents one third of the total resistance and would have one third (0.5 volt) of the total voltage across it. A sensitive voltmeter would record this drop.

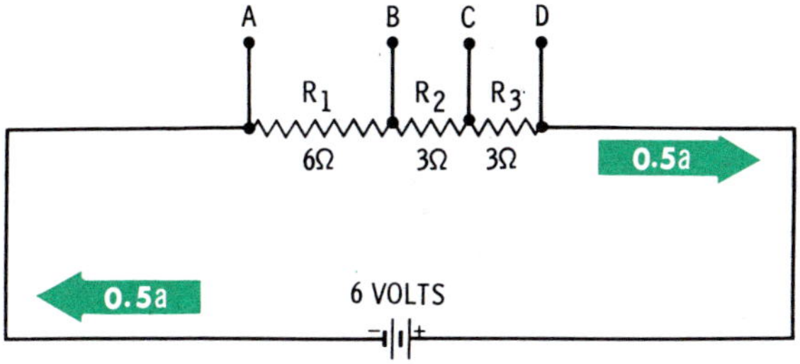


**Voltage drop in a load.**

Normally, the relatively small voltage drop of conductors is disregarded since the wire resistance is usually a very small fraction of the total load resistance. The tapped resistance in the diagram above has a total of 12 ohms with a total of 6 volts (from the source) applied across it. A current of 0.5 amp flows through each portion. The voltage (IR) drop between terminals A and B will be 3 volts in accordance with Ohm's law, resistance ratio, or a voltmeter measurement.

**Q24. What is the voltage drop across R3?**

**Q25. What is the IR drop between terminals A and C?**

**Q26. How much voltage will be measured between terminals B and D?**

**Q27. The sum of the total voltage drops in a circuit is (equal to, less than, more than) source E.**

**Q28. R5 is 0.2 of the total circuit resistance. The circuit voltage is 9 volts. What is the IR drop across R5?**

**Your Answers Should Be:**

**A24.** The voltage drop across R3 is **1.5 volts.**

**A25.** **4.5 volts** (0.5 amp × 9 ohms) is distributed between terminals A and C.

**A26.** **Three volts** between B and D.

**A27.** The sum of the total voltage drops in a circuit is **equal to** source E.

**A28.** **1.8 volts** (two tenths of nine volts.)

## ELECTRIC POWER

When voltage forces current through a resistance, heat is generated. **Electrical energy** is converted to **heat energy.** The rate at which this conversion takes place is called **power,** and its unit of measurement is the **watt.** Power is determined by the product of current flowing through the device and the voltage dropped across the device.

$$P \text{ (power in watts)} = E \text{ (volts)} \times I \text{ (amps)} = EI$$

Since P = EI, the wattage rating of a device reveals its voltage or current if one of these values is known. For example, an electric light bulb has its wattage and voltage stamped on its surface. Why do you think this is done?

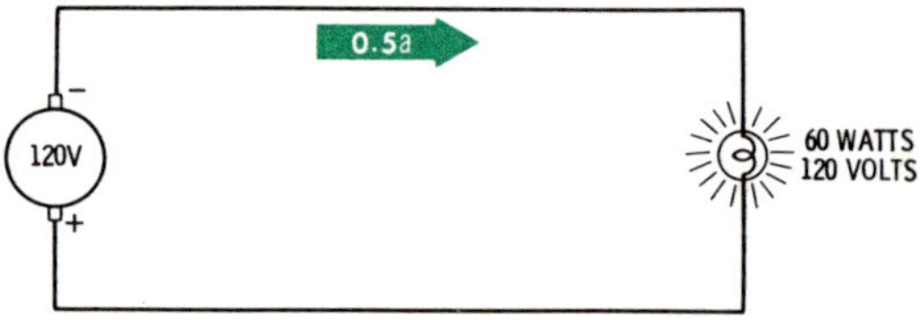


**Power dissipation in a lamp.**

Connected to 120 volts, the 60-watt lamp will draw 0.5 amp. If P = IE, I is equal to P divided by E. If the filament has been properly constructed, the lamp will burn for many hours with ½ ampere flowing through it. What will happen if the lamp is connected across 240 volts? Since the voltage is doubled, current will also double. The power (IE) that the lamp must now dissipate is 1 ampere multiplied by 240 volts, which is equal to 240 watts. The filament, constructed for 60 watts, will be rapidly consumed by the increased heat. For this reason electrical devices should be connected to proper voltages only.

Power can be determined by knowing only E and R, or only I and R. The power equations can be developed by substituting the appropriate Ohm's law equations in P = IE.

If P = IE, and E = IR, then (by substitution):
$P = (I)(IR) = I^2R.$

Since $I = \frac{E}{R}$, $P = \left(\frac{E}{R}\right)(E) = \frac{E^2}{R}$

By use of the appropriate expression, power can be determined if the voltage and current, the current and resistance, or the voltage and resistance are known.

Since electrical energy is dissipated in the form of heat in a resistance, the power developed in a resistance is considered to be a loss. If 1 ampere of current causes a voltage drop of 120 volts, the power (IE) loss is 120 watts. If 2 amps flow through 10 ohms, the power loss (also called an $I^2R$ loss) is 40 watts. A voltage drop of 10 volts across 20 ohms of resistance will dissipate 5 watts of power ($E^2/R$).

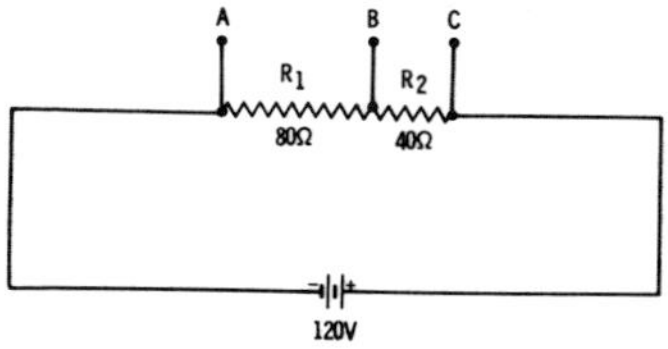


**Power losses in a circuit.**

**Q29. What is the total circuit current in the above circuit?**

**Q30. How much current is flowing through $R_1$?**

**Q31. What is the voltage drop between terminals A and C?**

**Q32. How many volts will be measured across $R_1$?**

**Q33. What is the IR drop across $R_2$?**

**Q34. What is the power loss in $R_1$?**

**Q35. How much power is dissipated in the total load?**

**Q36. If the source voltage is decreased to 60 volts, how many watts will be dissipated by $R_2$?**

**Your Answers Should Be:**

**A29.** **1 ampere.** (Source voltage of 120 volts divided by the total resistance of 120 ohms.)

**A30.** **1 ampere.** (The same current flows through all parts of this circuit.)

**A31.** **120 volts.** ($E = 1 \text{ amp} \times 120 \text{ ohms}$)

**A32.** **80 volts.** ($E = IR$)

**A33.** **40 volts.**

**A34.** **80 watts.**

**A35.** **120 watts.**

**A36.** **10 watts.** This can be determined by several methods based on the current now being 0.5 amp ($E/R$) and the new voltage drop of 20 volts across $R_2$:

(a) $P = IE = (0.5)(20) = 10 \text{ watts}$

(b) $P = \frac{E^2}{R} = \frac{(20)^2}{40} = \frac{400}{40} = 10 \text{ watts}$

(c) $P = I^2R = (0.5)^2 \times (40)$
$= (0.25)(40) = 10 \text{ watts}$

(Note that decreasing the voltage to one half the original value halved the amount of current flowing and decreased the power consumption of the circuit to one fourth.)

As in the Ohm's law equations, all factors in the power equations must be in equivalent units. E must be in volts; I in amperes; R in ohms.

## WHAT YOU HAVE LEARNED

1. Current will flow only in a complete circuit.
2. A circuit consists of a source connected to a load.
3. A basic circuit consists of a source, conductors, and a load.
4. Switches are used to open and close the current path.
5. A toggle switch is the most common switch.
6. A knife switch was probably the first form of switch.

7. Wafer switches rotate and may have more than one set of terminals.
8. A push-button switch is designed to be either normally open or normally closed. Depressing the switch button opens a closed circuit or closes an open circuit.
9. Current will not flow through an open circuit but will flow in a closed circuit.
10. Because of the Ohm's law relationship between I (current), E (voltage), and R (resistance), the value of one of these factors can be determined if the other two are known.
11. The voltage applied across a resistance is equal to the current flowing through the resistor times its resistance. $E = I \times R$.
12. Current through a resistor is equal to the voltage applied divided by the resistance:

$$I = \frac{E}{R}$$

The greater the resistance, the smaller the current will be for a given voltage.
13. Resistance is equal to the amount of voltage applied divided by the amount of current the resistor permits to flow:

$$R = \frac{E}{I}$$

14. The power dissipated in an electrical circuit generates heat. The amount of heat depends on how much current is flowing and the amount of voltage forcing it to flow.
15. $P = IE$, where the power is measured in watts (w).
16. Power is equal to $I^2R$, where IR is substituted for E in the $P = IE$ expression.
17. Power is equal to $E^2/R$, where $E/R$ is substituted for I in the $P = IE$ expression.
18. Ohm's law expressions:

$$E = IR,\ I = \frac{E}{R},\ R = \frac{E}{I}.$$

19. Power expressions:

$$P = IE,\ P = I^2R,\ P = \frac{E^2}{R}.$$

# 3

# DC Series Circuits

## *What You Will Learn*

**This chapter contains a thorough description of the series circuit and its basic connections. It also explains how total resistance, total current, and voltage drops are determined in such a circuit. You will learn how to reduce a voltage to a desired level by the use of a dropping resistor, identify series connections in complex electronic circuits, and determine total resistance and total current in a series circuit.**

### WHAT IS A SERIES CIRCUIT?

A **series circuit** is an electrical circuit in which all the components are connected end to end.

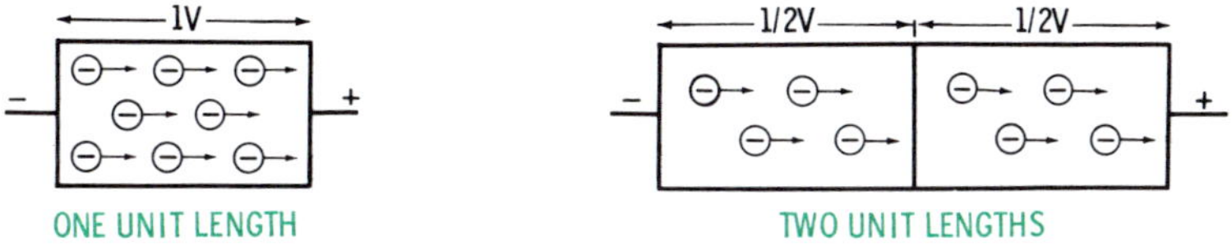


**Voltage and current.**

Do you recall the voltage distribution which occurs across a resistance? If one volt is applied across one unit length of wire, for example, what will happen to the voltage distribution across the same size wire that is twice as long? One-half volt will appear across each unit length. One-half the total voltage will be dropped across each of the two units. Since the resistance is doubled, the current will be one-half as much.

## VOLTAGE DISTRIBUTION

Voltage will be distributed across the unit length of resistance in the manner shown in the diagram. Doubling the voltage will cause twice the current to flow through the

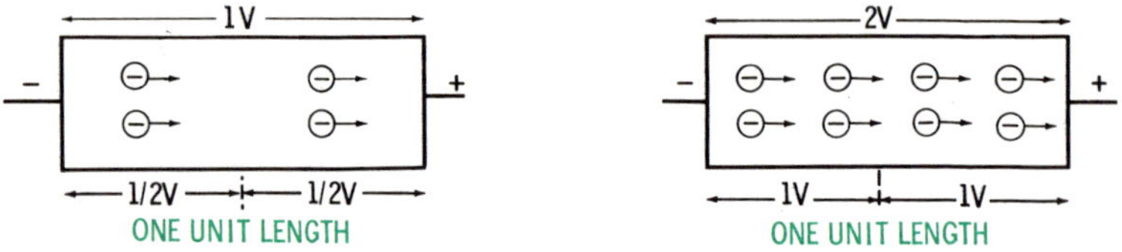


**Current depends on voltage.**

resistance. One volt is distributed (dropped) across each half of the resistance, increasing the current that flows through its section.

An example of a series circuit is the manner in which the vacuum-tube filaments of some radio and television sets are connected. As you can see in the diagram below, each filament requires 12 volts and a current of 0.15 ampere. This identifies another characteristic of a series circuit—all components in a series circuit have the same current flowing through them.

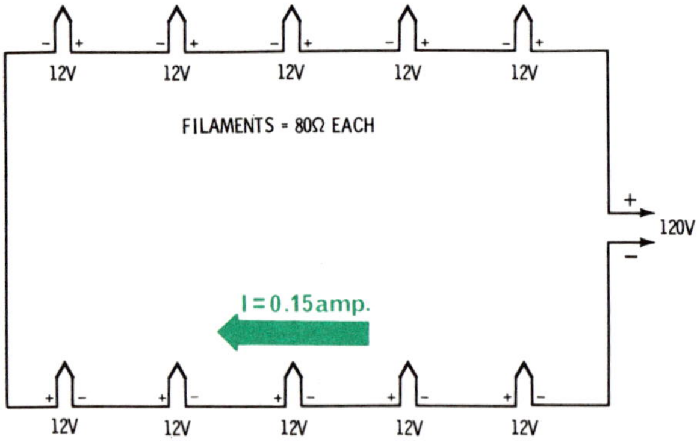


**Vacuum-tube filaments in series.**

Another application of the series circuit is the economical series-string Christmas-tree lamps. The number of lamps needed in a string or the amount of voltage to apply can be determined by Ohm's law. If the voltage drop required by each lamp is 15 volts and the string is to be connected to a 120-volt outlet, then eight lamps are required in series.

### Symbol Designations

Care must be used when referring to voltage in a series circuit. A voltage drop across one resistance among many may be a different value from the IR drop across the others. A voltage across $R_1$, for example, should be identified as that voltage. The source voltage should be designated as $E_T$ (for E total). Total resistance becomes $R_T$. Since current is the same in all parts of a series circuit, it remains as I.

**Identify all parts of a circuit.**

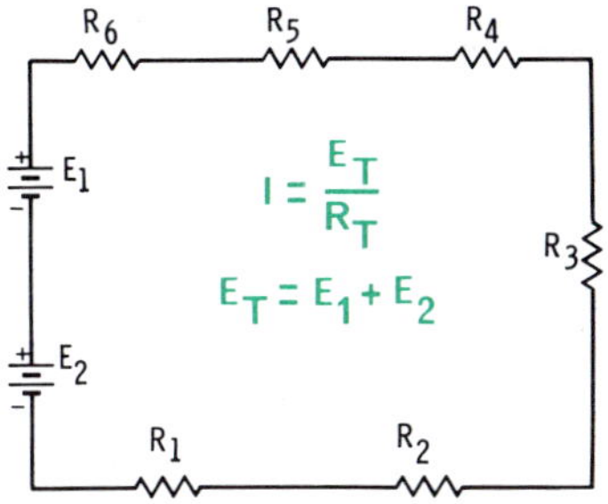


### Total Resistance in the Series Circuit

Current in a series circuit is determined by the values of **total resistance** and **total voltage.** The total source voltage is distributed proportionally across each of the series resistances, depending on their ratios to the total resistance.

Total resistance in a series circuit is the sum of the resistances between the terminals of the source. That is, $R_T$ will equal $R_1 + R_2 + R_3 +$ etc.

**Total resistance.**

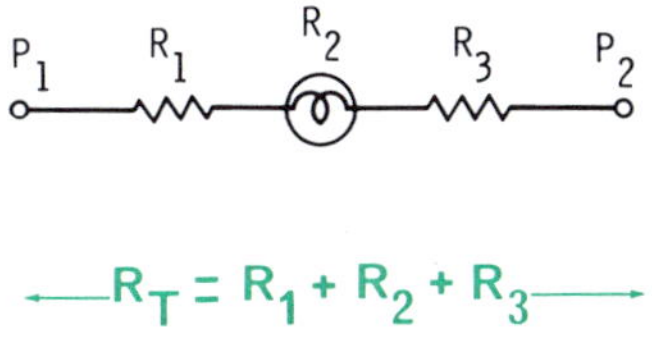


The lamp shown above does have a resistance, even though it is not a resistor. Therefore it is marked as $R_2$ for calculation purposes.

**Q1. Can the circuit on the opposite page be called a series circuit?**

**Q2. The symbol for total voltage is ______.**

**Q3. If $R_1$ is 4.5 ohms, $R_2$ is 6 ohms, and $R_3$ is 6 ohms in the diagram above, what is $R_T$?**

> **Your Answers Should Be:**
>
> **A1. Yes.** All elements are connected end to end.
>
> **A2.** The symbol for total voltage is $\mathbf{E_T}$.
>
> **A3.** $\mathbf{R_T = 16.5}$ **ohms.** (The total resistance in a series circuit is the sum of all the resistance values across the source.)

## Total Voltage in a Series Circuit

In addition to determining the total resistance of a series circuit, the total voltage must be calculated if there are two or more voltage sources in series. Total voltage is found in the same manner as the total resistance; that is, the sum of the individual voltages equals the total voltage.

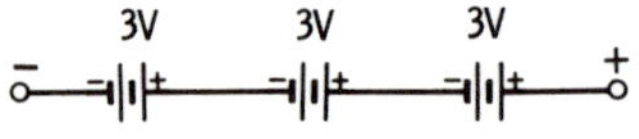


**Total voltage.**

$$\leftarrow E_T = E_1 + E_2 + E_3 = 9V \rightarrow$$

There is a separate problem in the calculation of total voltage, however. Some of the voltage sources may be in opposition to others, in which case the total voltage will not be the simple numerical sum of all the voltages. If the polarities of all the voltage sources are in the same direction, the voltage values are added together. If the polarities are in opposite directions, the values are subtracted, and the polarity of $E_T$ is that of the larger voltage source.

A series circuit with more than one source and more than one load can be redrawn to show a circuit with only one source and one load.

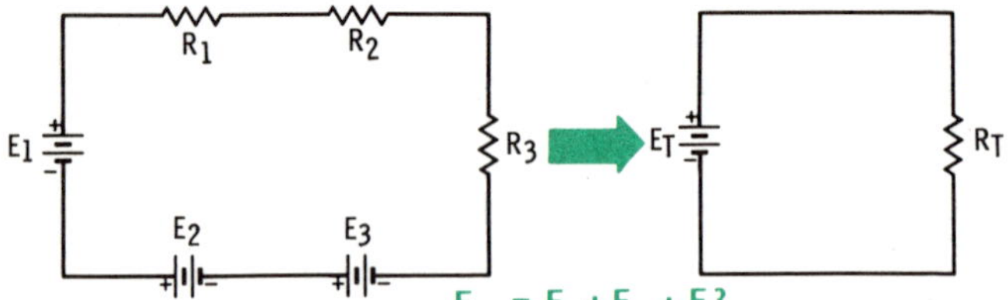


$$E_T = E_1 + E_2 + E3$$

$$R_T = R_1 + R_2 + R_3$$

**Simplifying a series circuit.**

Q4. How must you treat the following series circuit to find the total source voltage?

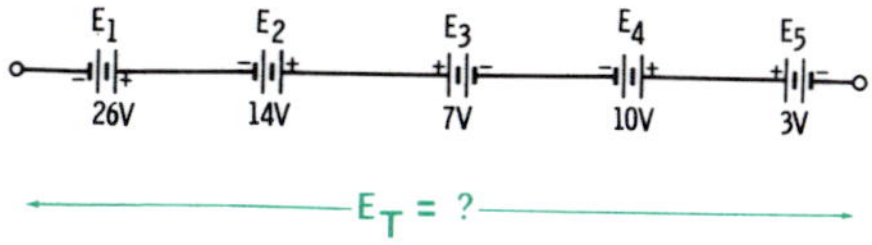


Q5. What is $E_T$ for the circuit below?

Q6. What is $R_T$ for the circuit below?

Q7. What is $I_T$ for the circuit below?

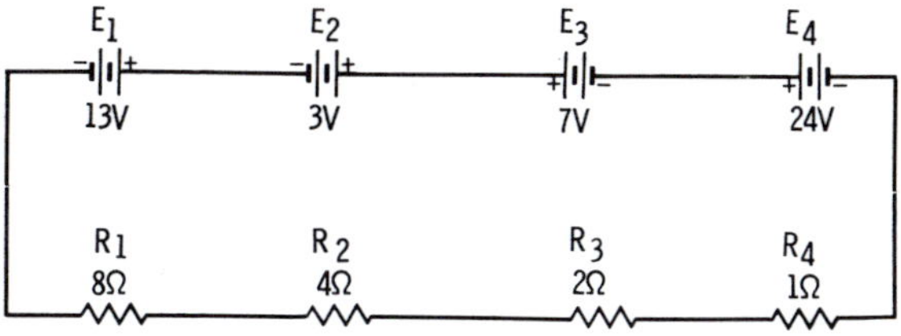


Q8. In which direction will current flow in the circuit above?

Q9. Could the total resistance be represented by a single resistor?

Q10. What is the total resistance in each circuit (a, b, and c) below?

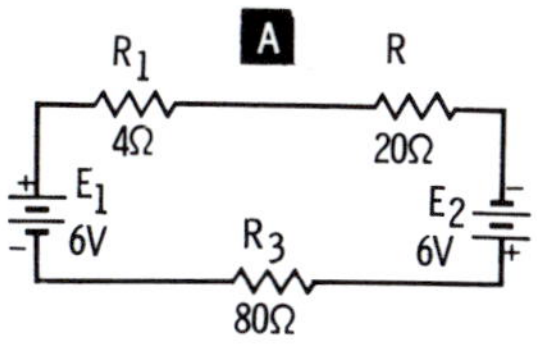


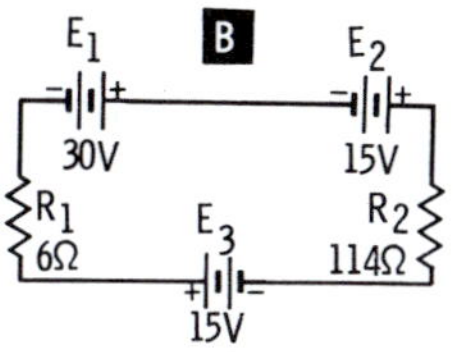


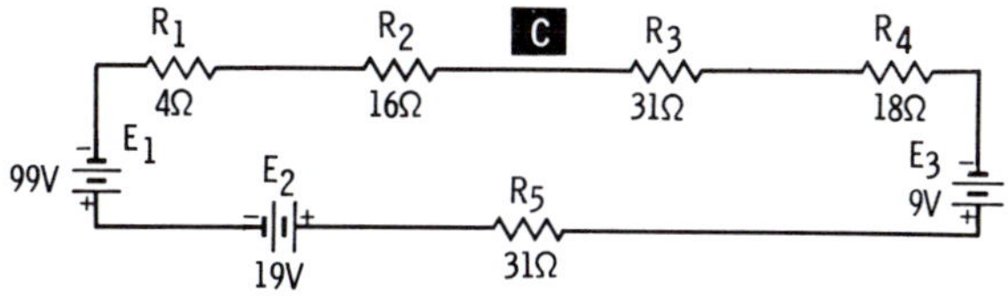


Q11. What is the total voltage in each circuit?

Q12. Draw an equivalent circuit, containing only one source and one load, for each of the circuits. Label the value and polarity of the source and the value of the resistance.

**Your Answers Should Be:**

**A4.** You must first add all the voltages having a polarity in one direction (negative to plus). Then add all the other voltages having a polarity in the other direction. Subtract the smaller from the larger. This will determine the amount of the overall voltage ($E_T$) and its polarity.

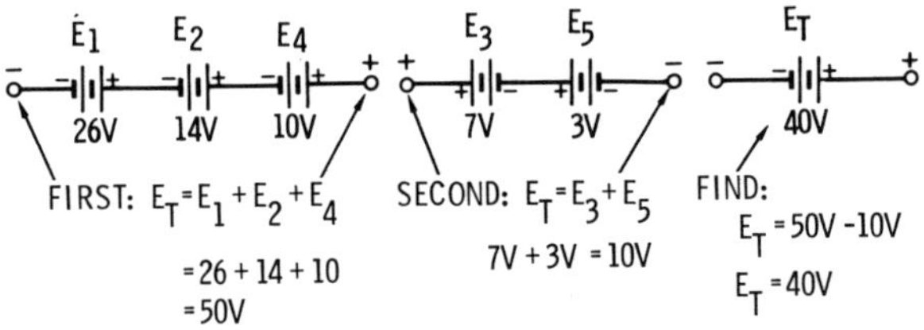


**A5.** $E_T = 24 + 7 - 13 - 3$. That is, $31 - 16 = 15$. Therefore, $\mathbf{E_T = 15}$ **volts,** negative to positive right to left.

**A6.** $R_T = R_1 + R_2 + R_3 + R_4$. $\mathbf{R_T = 15}$ **ohms.**

**A7.** $I_T = E_T/R_T$. $I_T = 15/15$, or **1 amp.**

**A8.** The current will flow **away from the negative terminal and into the positive terminal of** $\mathbf{E_T}$**.**

**A9.** **Yes.** Once the total resistance has been found, it may be represented by a single resistor.

**A10.** (a) **104 ohms**
(b) **120 ohms**
(c) **100 ohms**

**A11.** (a) **12 volts**
(b) **60 volts**
(c) **109 volts**

**A12.**

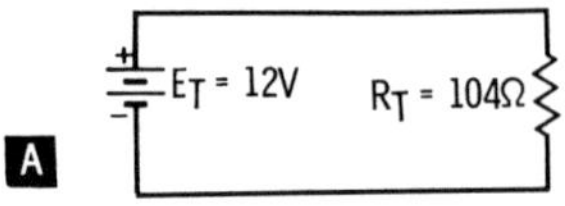


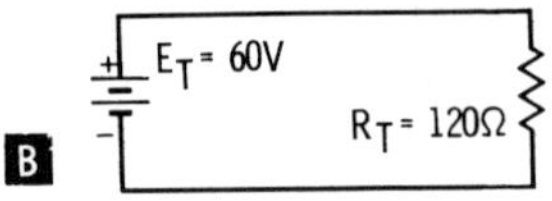


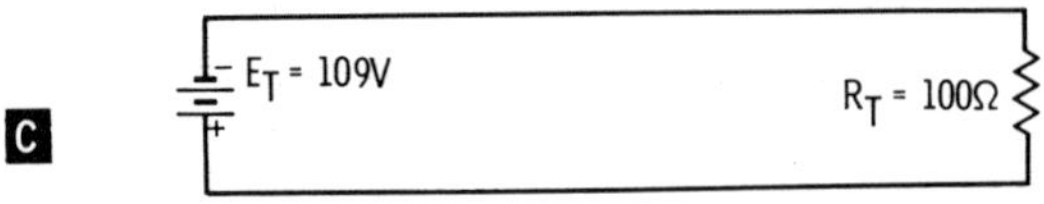

Examples of devices having several voltage sources in series are flashlights, transistor radios, and automobile batteries.

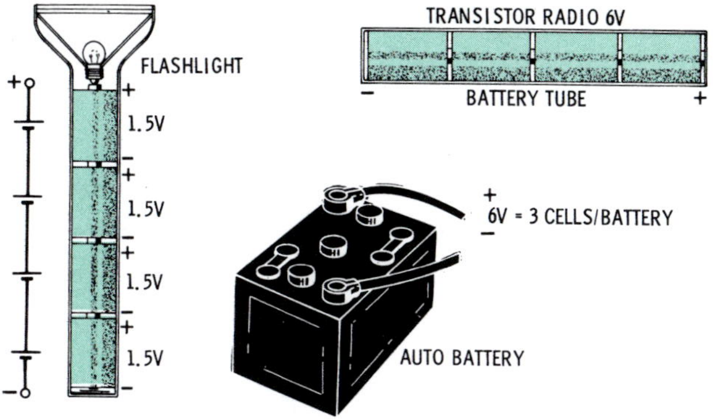


**Voltage sources in series.**

## Current in a Series Circuit

If a series circuit has an $R_T$ of 170 ohms and an $E_T$ of 34 volts, what is the current in the circuit? The current will be equal to $E_T/R_T$ or 0.2 ampere (34/170). If a series circuit has three sources (all aiding one another) and two resistances, current can be determined by:

$$I_T = \frac{E_1 + E_2 + E_3}{R_1 + R_2}$$

## Voltage Drop in a Series Circuit

The voltage drop across each resistance in a series circuit is found in the same manner as the voltage across a resistor if the resistance value and the current flowing through the resistor are known. The Ohm's law expression is $E = IR$. Voltage across a resistance is determined by multiplying current by resistance.

The current is the same through each resistance in a series circuit. After finding the current, the value of I is used to determine the voltage drop across each resistance in the circuit.

**Q13. What is the total voltage of a source having eight 1.5-volt flashlight cells connected in series?**

**Q14. What is the total voltage of a source having fifty-five 2-volt lead-acid cells connected in series?**

**Your Answers Should Be:**

**A13.** Eight 1.5-volt dry cells connected in series will develop **12 volts.**

**A14.** Fifty-five 2-volt lead-acid cells connected in series will provide **110 volts.**

### Determining Voltage Drops

The sum of the voltage drops in a series circuit is always equal to the total applied voltage.

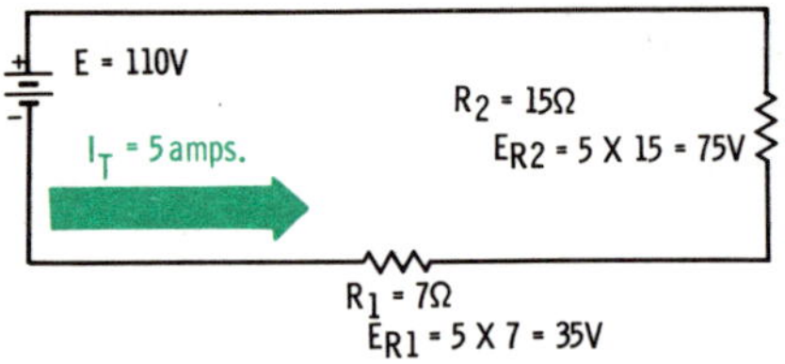


**Voltage drops equal applied voltage.**

In the above circuit, 5 amps will cause a 35-volt drop across the 7-ohm resistor and a 75-volt drop across the 15-ohm resistor. $E_{R1} + E_{R2} = E_T$.

### Polarity Across the Loads

Current leaves the negative terminal of a voltage source, flows through the circuit, and returns to the positive terminal. This direction of current flow occurs because of the voltage polarity. One of the terminals of the source is negative (repels electrons) with respect to the other. The opposite terminal is positive (attracts electrons). In the diagram, for example, $P_1$ is 30 volts negative with respect to $P_3$.

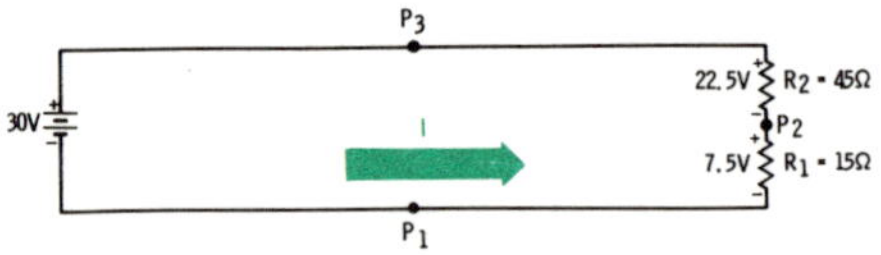


**Polarity across loads.**

Voltage in a circuit exists only between two points—never at one point only. Therefore, voltage is expressed as being across two points in a circuit, or in terms of one point with respect or in reference to another point. The point at which current enters a resistance is negative with respect to the point at which it leaves.

A **common ground** is used as the reference point for expressing voltages unless otherwise specified. If a ground symbol is shown (see the diagram below), it becomes the common reference point for all voltage points in the circuit.

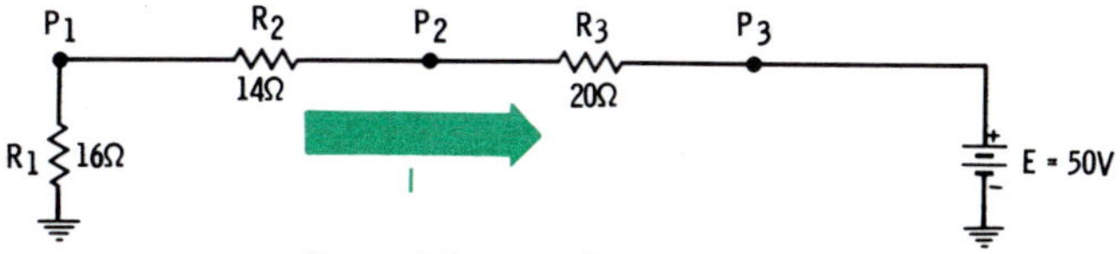


**Ground is a reference point.**

Current will flow from $P_1$ to $P_2$ to $P_3$ through the source to ground and from ground to $P_1$. In this case, I is equal to 1 amp. This means $P_1$ is +16 volts with respect to ground. $P_2$ is +14 volts with respect to $P_1$ (thus 30 volts positive to ground). $P_3$ to ground (in either direction) is +50 volts (the source voltage or the drop across the three resistors).

**Q15. What is the value of $E_T$ in the circuit below?**

**Q16. What is $I_T$ in the following circuit? Which way will current flow with respect to $P_1$ and $P_2$?**

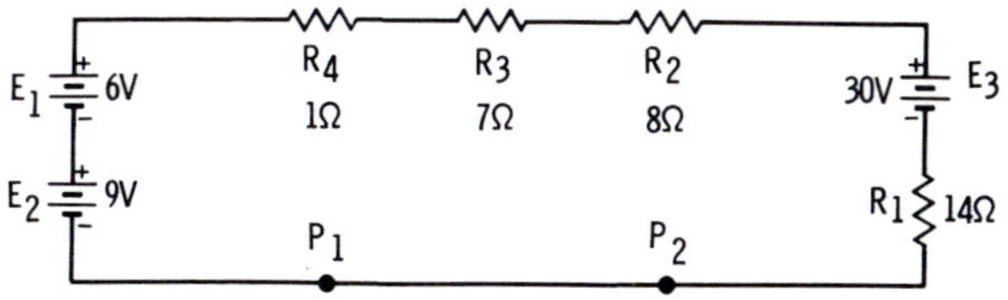


**Q17. What is the total resistance of the following circuit?**

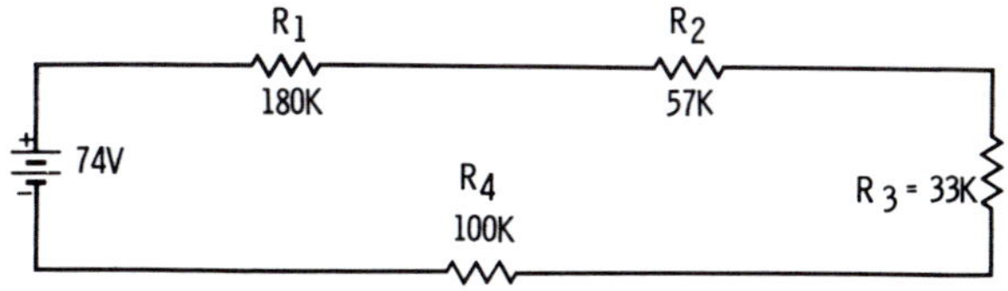


**Q18. What is the voltage, with reference to ground, for all the points (P) on the following diagram?**

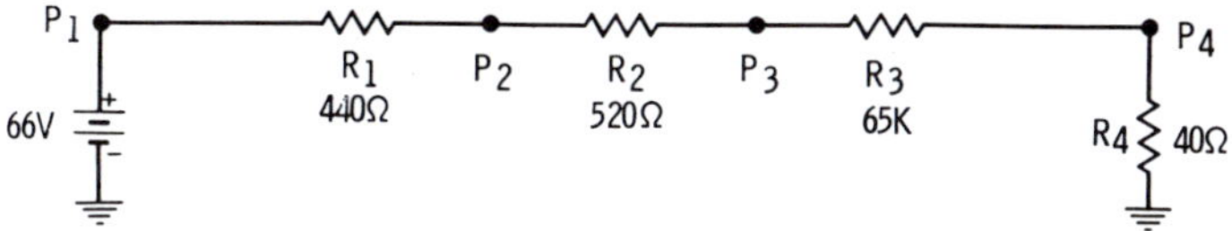

**Your Answers Should Be:**

**A15. 15 volts.**

**A16.** $I_T = E_T/R_T$. $E_T = E_3 - (E_1 + E_2) = 15V$

$R_T = R_1 + R_2 + R_3 + R_4 = 30$ ohms

$I_T = 15V/30 =$ **0.5 amp**

Since the 30-volt source is larger than the combined 15 volts of the other two, the **current will flow from $P_2$ toward $P_1$.**

**A17. $R_T = 370K$.**

That is, $R_T = R_1 + R_2 + R_3 + R_4 = 370K$

**A18.** $I_T = \dfrac{E_T}{R_T} = \dfrac{66V}{66K} = 0.001$ amp, or 1 ma

**$P_1 = +66V$, $P_2 = +65.56V$,**
**$P_3 = +65.04V$, $P_4 = +0.04V$**

## Voltage Division in a Series Circuit

Voltage reference is one of the most important concepts to be learned in electricity or electronics. An understanding of how much voltage exists between two points in a circuit often reveals the purpose of the circuit and how it works.

As an example, a schematic of a vacuum-tube circuit is shown below. The figure on the right shows a vacuum tube in series with a load resistance, $R_L$. The figure on the left shows how the tube can be considered as a variable resistance. The output of this circuit is taken at a point between the tube and the load resistance.

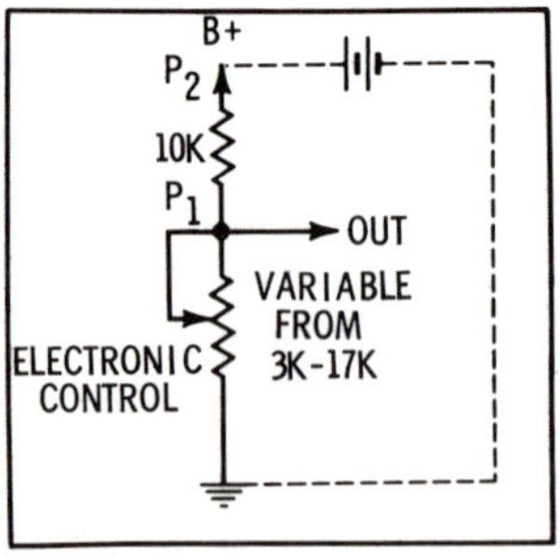


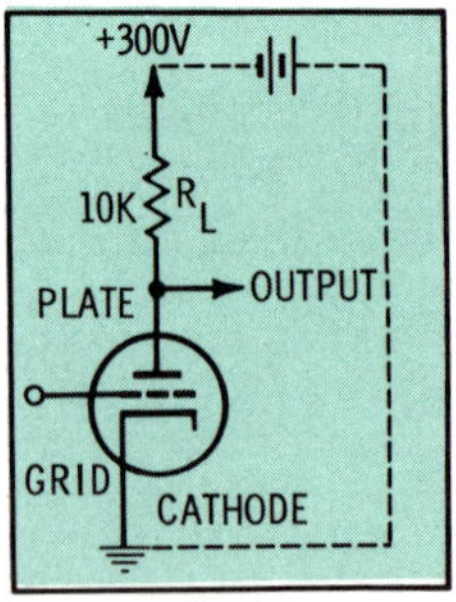


In effect, the grid varies the resistance of the vacuum tube. $P_2$ is always +300V (to ground). $P_1$ is a positive

voltage (to ground); its value depends on tube resistance at any particular instant. The tube resistance changes from 3,000 to 17,000 ohms. Since 300 volts is always across the tube and load resistance, the output voltage at $P_1$ (with respect to ground) is large when tube resistance is high and small when tube resistance is low.

## VOLTAGE DIVIDER

Current is the same through all elements in a series circuit. In the circuit on the opposite page the 300 volts can also be used for other purposes and in other circuits. Yet, 300 volts is often too much voltage for some circuits. A voltage divider is used to reduce the 300 volts to a level acceptable for use in a lower-voltage circuit. To select the correct resistors for use in a voltage divider, you must know how much voltage is required by the lower-voltage circuit. A voltage divider, therefore, is a series of resistances whose values are such that the desired output voltages are obtained at the various points with respect to the voltage reference point. In some cases, a voltage dropping resistor is connected in series with a load to obtain the desired voltage.

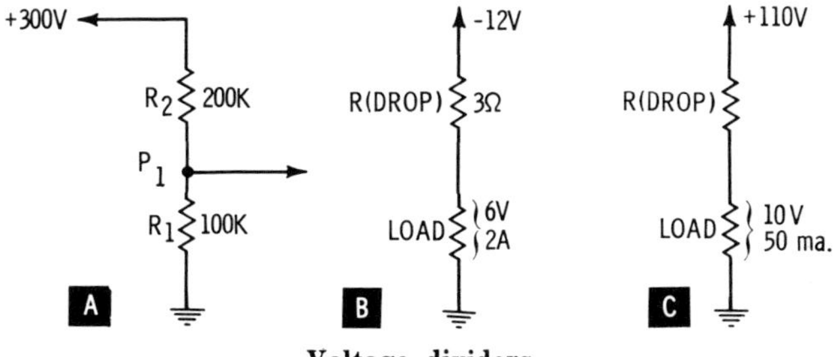


**Voltage dividers.**

**Q19. What is I in the voltage divider of part A above?**

**Q20. How much voltage is available at $P_1$ in part A?**

**Q21. R (drop) in part B above is a(an) ________________ resistor.**

**Q22. Load voltage (part B) is ______ to ground.**

**Q23. ___ ampere(s) flow through R (drop) in part B.**

**Q24. R (drop), part C, is ______ ohms.**

**Q25. The load in part C dissipates ______ watt(s).**

**Your Answers Should Be:**

**A19.** **0.001 amp**, or 1 ma.

**A20.** **100 volts.** $E = I \times R_1$. 0.001 amp multiplied by 100,000 ohms.

**A21.** R (drop) in part B is a **voltage dropping** resistor.

**A22.** Load voltage (part B) is **—6V** to ground.

**A23.** **Two** amps flow through R (drop) in part B. (Current is the same throughout a series circuit.)

**A24.** R (drop), part C, is **2,000** (or 2K) ohms. If the voltage across the load is 10 volts, R (drop) must have 100 volts across it. 100 volts divided by 0.05 amp (circuit current) is 2,000 ohms.

**A25.** The load in part C dissipates **0.5** watt. $P = IE$. (Current through the load multiplied by the voltage across the load.)

## PRACTICAL APPLICATION OF THE SERIES CIRCUIT

In addition to what you have learned, there are many other applications for a series circuit. They exist in almost every electrical device used.

### Reducing Output Voltage of a Battery

A voltage-dropping resistance can be used to lower the output of a 12-volt car battery to operate a 6-volt device (radio, meter, lamp, etc.). The dropping resistor (when in series with the load) must have a value in ohms which will

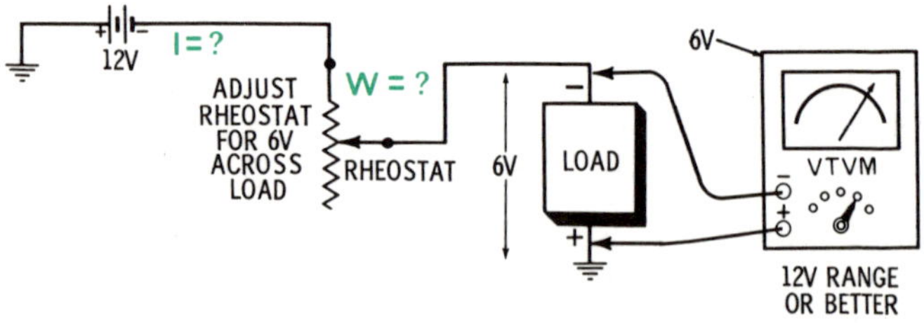


**Reducing voltage with a rheostat.**

permit the desired amount of current to flow through the device. It must also have the proper wattage rating as determined by its current and voltage drop. A rheostat (variable resistor) is quite often used as a dropping resistor in this application.

The circuit on the opposite page requires many careful adjustments and a common-sense application of Ohm's law and power equations. Solutions to the voltage-current-resistance problems are no more difficult than those you have already solved.

**Steps in the Adjustments of the Rheostat**

1. Determine the approximate current the load will draw. The device should be marked with its voltage and current requirements. The information can also be found in the service manual for this particular component.
2. Be sure the multimeter is set on a voltage scale which will read the source voltage (12 volts in this case).
3. Adjust the rheostat for 6 volts across the load.

   **NOTE:** The rheostat must be of high enough wattage to carry the load current.
4. Be sure the device can be switched on and off.
5. Put a fuse in the device. The fuse should be capable of carrying the current requirements of the load, while protecting it from an accidental overload.

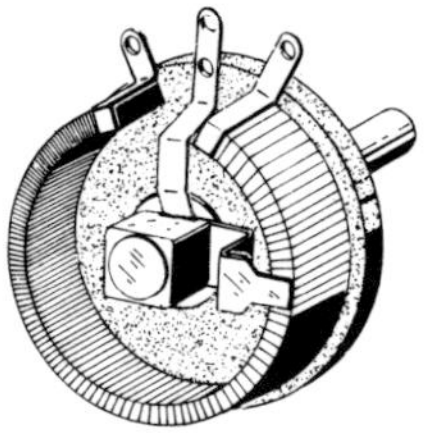

*Rheostat.*

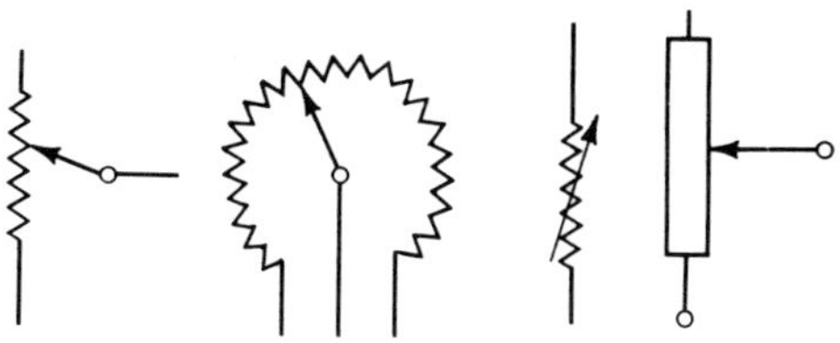

*Rheostat symbols.*

**Rheostat and symbols.**

**Q26. If you do not know the current through the load and the rheostat, and you do not have an ammeter which will measure the current, how do you find the power required for the rheostat?**

**Q27. How do you determine the size of the fuse?**

**Q28. Why should the multimeter be set to the range which can read the voltage of the source?**

**Your Answers Should Be:**

**A26.** You know the voltage drop across the rheostat, and you can measure the resistance of the rheostat from the end of the wiper contact.

$$\text{Power} = \frac{E^2}{R}$$

**A27.** The fuse must be able to carry the calculated current and should be able to carry normal surges. Normal surges are determined by the characteristics of the device (load).

**A28.** The voltage range of the multimeter must always be set to the scale that you know will not be exceeded. Voltage greater than the range of the meter can cause excess current through the meter and, therefore, cause possible damage.

## Speed Control for an Electric Motor

Motors, such as the one used in an automobile for the defroster fan, are sometimes connected to a switch having four positions (OFF-LO-MED-HI). The low and medium positions are separated from the high position by fixed resistors. Assume the motor turns at three speeds which require 3V at 1 amp (LO), 4.5V at 1.5 amps (MED), and 6V at 2 amps (HI). This circuit is shown below.

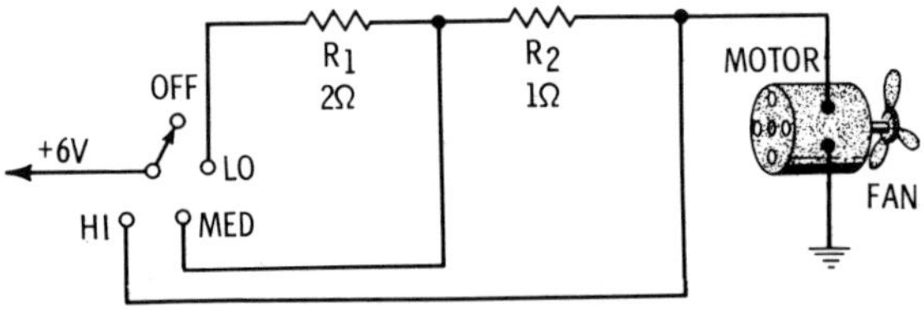


**Speed-control circuit.**

In the LO position the series resistance equals 3 ohms, which permits 1 ampere of current to flow. The 3 ohms is not the only resistance in the circuit. The motor adds its resistance in series with the circuit. In the MED position of the switch, the series resistance is 1 ohm. Again, the motor resistance is in series with it.

Q29. What is the resistance of the motor when the switch is in the LO position?

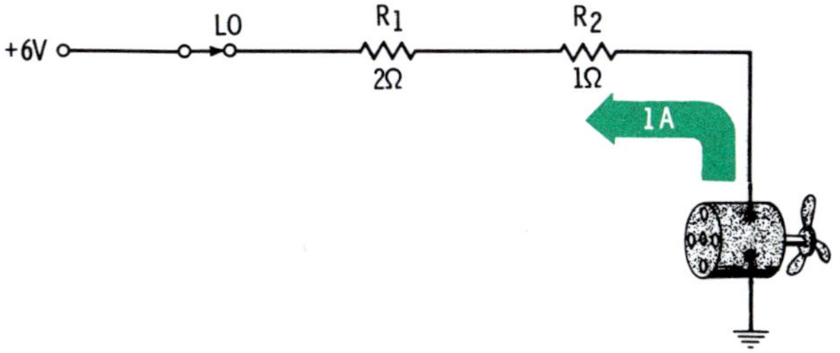


Q30. What is the resistance of the motor when the switch is in the MED position?

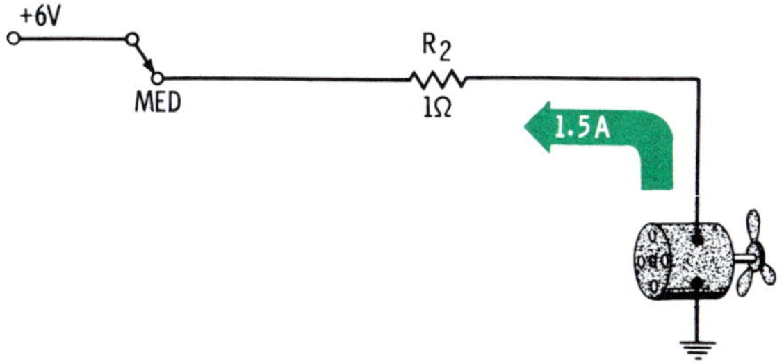


Q31. What is the resistance of the motor when the switch is in the HI position?

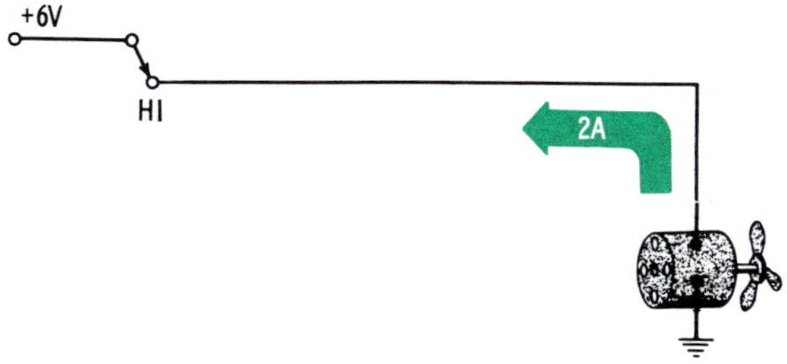


Q32. What is the power requirement for each of the two resistors, $R_1$ and $R_2$?

Q33. What type of switch would be the one most likely used?

**Your Answers Should Be:**

**A29.** The voltage applied to the motor in the LO position is 3V. The current is 1 amp. Therefore:

$$R = \frac{E}{I} = \frac{3}{1} = \mathbf{3\ ohms}$$

**A30.** The voltage drop across the dropping resistor is equal to 1 ohm times 1.5 amps. This 1.5V drop across the 1-ohm resistor is subtracted from the source ($E_T$).

$$E_{motor} = E_T - E_{R2} = 4.5V$$

$$R_{motor} = \frac{E_{motor}}{I_T}$$

$$R = \frac{4.5V}{1.5A} = \mathbf{3\ ohms}$$

**A31.** **3 ohms again.** The full 6 volts is across the motor.

**A32.** It is determined by the largest amount of current which will flow through the resistors.

$P_{R2} = I_{R2} \times E_{R2} = 1.5 \times 1.5 = 2.25$ watts

$P_T = (E_{R1} + E_{R2}) \times I_T$. $E_{R1} = R_1 \times I_T$, and $E_{R2} = R_2 \times I_T$.

$E_{R1} = 2 \times 1 = 2V$, $E_{R2} = 1 \times 1 = 1V$.
$P_{R1} = 1\ amp \times 2V = 2$ watts, $P_{R2} = 1$ watt.

Therefore, $R_1$ must have a power rating no smaller than **2 watts.** The rating of $R_2$ must be no smaller than **2.25 watts.**

**A33.** The most probable selection for the switch would be a **rotary or wafer type.**

## WHAT YOU HAVE LEARNED

1. The basic electrical circuit is a series circuit.
2. A series circuit may have more than one source and load.
3. Current in a series circuit is the same through all components.

4. Total resistance in a series circuit is computed by finding the sum of all the resistances. That is, $R_T = R_1 + R_2 + R_3$ plus whatever additional resistors may be in series.
5. Total source voltage in a series circuit is the sum of the individual sources if their polarity direction is the same, or is the difference of the sums of the opposing potentials. The larger will become $E_T$ and will control the direction of current.
6. A series circuit may be represented with an equivalent circuit.
7. Total voltage drop in a series circuit is equal to the source voltage, or $E_T$.
8. To describe a voltage at any given point you must identify its polarity with respect to a reference point.
9. A voltage divider is a series circuit which employs a dropping resistor to provide a desired voltage output.
10. When a dropping resistor is used to form a voltage divider, it must have a safe power rating.
11. In all practical applications of a dropping resistor or voltage divider, voltage, current, and power requirements must be calculated to determine the proper value and rating of the required resistor.
12. Symbols are used to represent components in schematic diagrams.

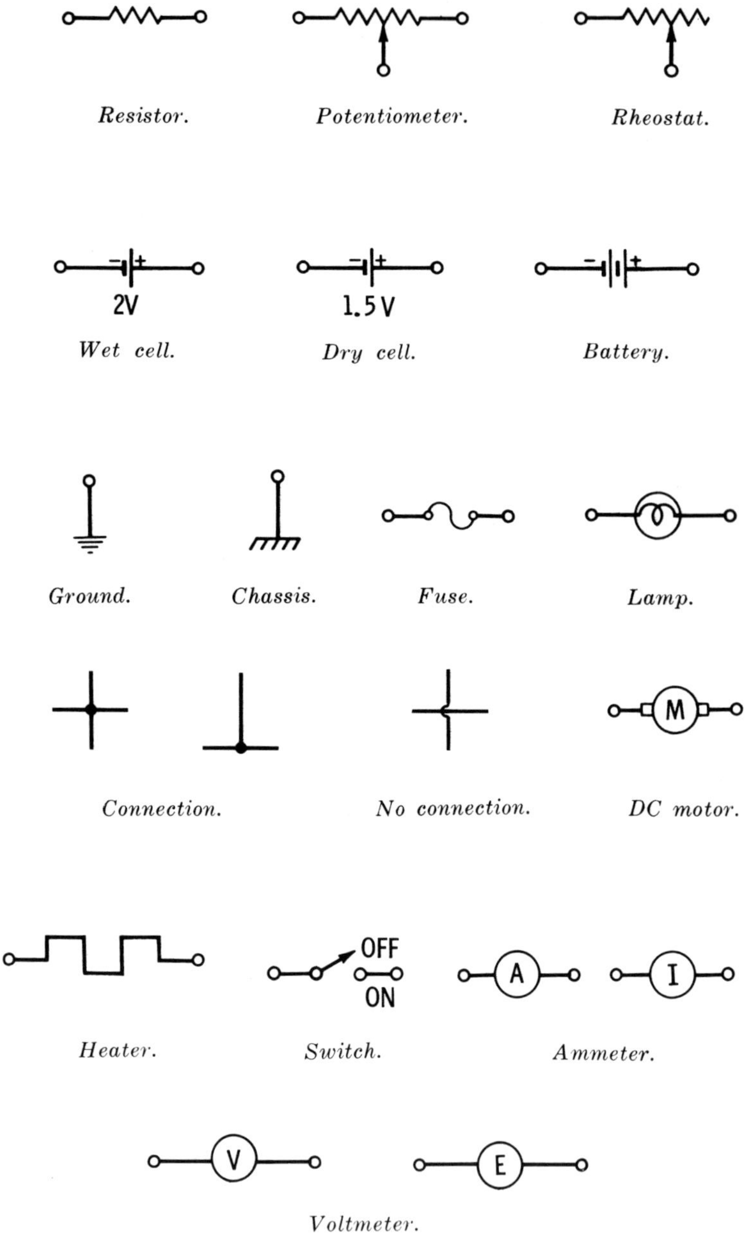


**Component symbols.**

# 4

# DC Parallel Circuits

## *What You Will Learn*

**Contained here is a thorough description of a DC parallel circuit. Included are the basic parallel connections, determination of total current and total resistance, the series equivalent circuit, current and voltage relationships, and typical applications of parallel circuits. You will learn how to identify parallel circuit networks, determine total current and total resistance, and develop series equivalent circuits.**

### WHAT IS A PARALLEL CIRCUIT?

A **parallel circuit** contains two or more basic circuits, each of which is connected to common terminal points. Two or more resistances, for example, may be connected together across the same voltage source.

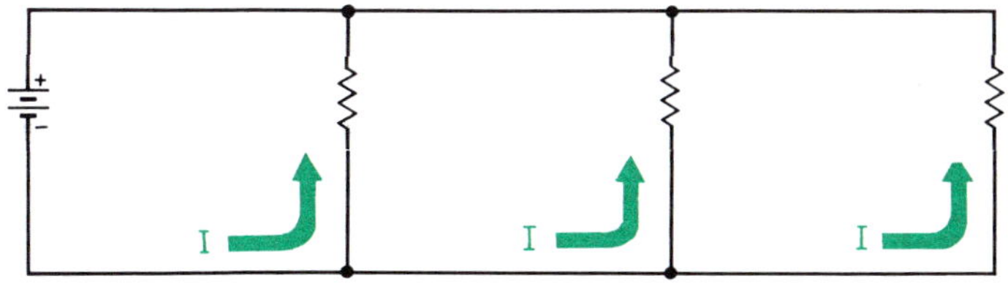


**A parallel circuit.**

A string of Christmas-tree lamps which will permit one or more lamps in the circuit to be opened and still allow the others to operate properly is one example of a parallel circuit. Each lamp has the same voltage applied to it.

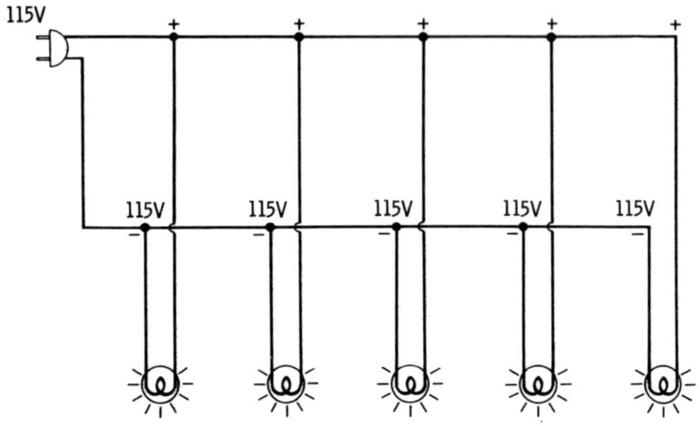


**Lamps in parallel.**

Another form of parallel circuit uses more than one voltage source in parallel to increase the availability of current. The greater the amount of stored energy available, the longer the source can produce a current at a given voltage.

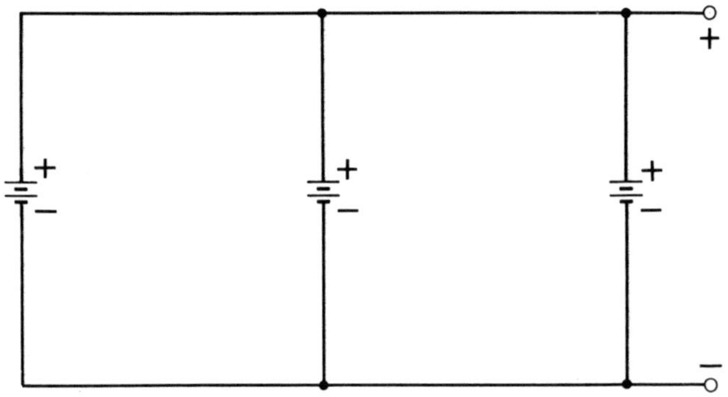


**More batteries—more available current.**

Still another example of a parallel circuit is the filament circuit for the vacuum tubes used in automobile radios. The filaments are all connected in parallel, and either 6 or 12 volts is applied to each. The tubes have either 6- or 12-volt filaments, depending on the type of battery in the automobile.

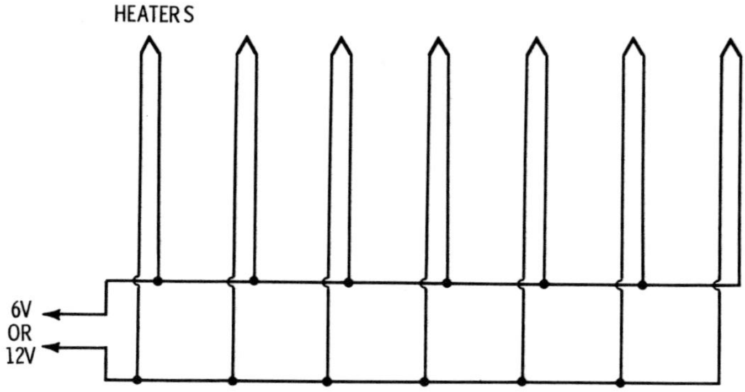


**Vacuum-tube filaments in parallel.**

In each case, you can see that all sources or all loads are connected across the same two points. This produces a circuit which has one common voltage applied to all loads.

You have previously learned that in a series circuit all loads and sources are connected end to end. The same current flows through all components, and the source voltage is divided among the separate loads. In a parallel circuit, all loads and sources are connected across the same points. Therefore, each load has the same voltage applied.

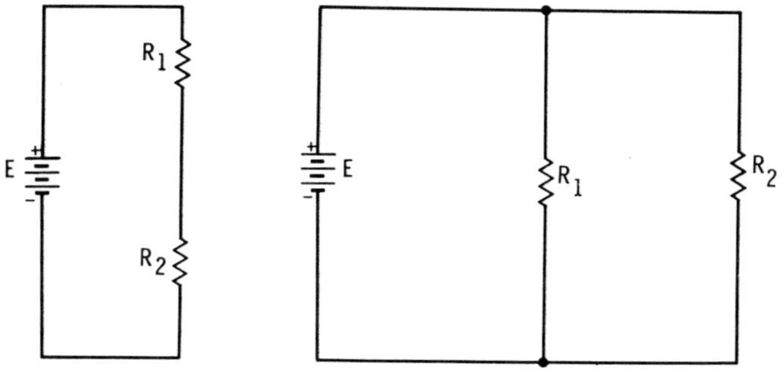


*Series circuit.* *Parallel circuit.*

**Load and source connections.**

**Q1. In a parallel circuit, each resistance can have the same (voltage, current) but different (voltage, current).**

**Q2. In a series circuit, each resistance can have the same (voltage, current) but different (voltage, current).**

**Your Answers Should Be:**

**A1.** In a parallel circuit, each resistance can have the same **voltage** but different **current.** With the same voltage across each load, current will equal the voltage divided by the load resistance.

**A2.** In a series circuit each resistance can have the same **current** but different **voltage.**

## AUTOMOBILE CIRCUITS

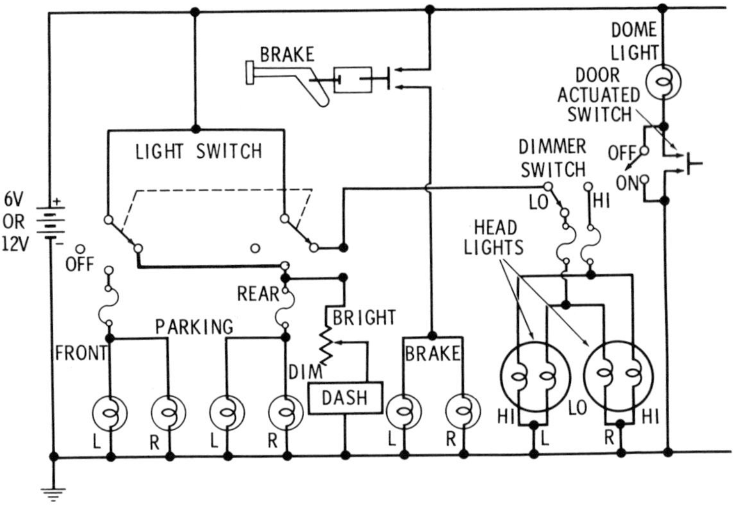


**Automobile light circuit.**

The diagram above is an example of a schematic for a lighting circuit in an automobile, and is also an example of a parallel circuit. Notice that fuses are used in the separate branches. This permits one circuit to short and blow its fuse while the other circuits remain energized. Notice that different switches are used for different jobs. The main light switch might be a rotary type (wafer). The stop-light switch is usually actuated by the same hydraulic fluid pressure that applies the brakes. That is, the switch is a push-button type that is spring-loaded to form a normally open circuit. When the brakes are applied, pressure builds up in the master brake cylinder, causing the stop-light switch to close. The head-light dimmer switch is located on the floor of most cars. It is a push-button switch which remains in one position until pressed again. The dome light is switched on and off by either of two switches in parallel. One is a push-button type that is spring-loaded to stay normally closed, and is actuated by opening and closing the car door. When the door opens, the switch returns to its normally closed position; when the door is closed, it pushes the switch open. In parallel with this switch is a slide switch, sometimes a part of the dome-light fixture.

**Slide switch.**

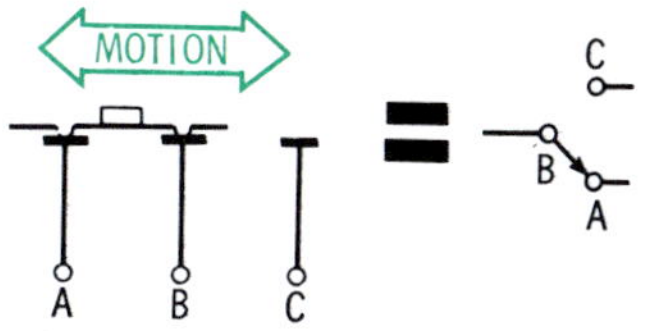


## CURRENT FLOW IN A PARALLEL CIRCUIT

Current in each branch of a parallel circuit must originate from the same source. This means that each branch will have a different current if the resistance of each branch is different. The source must supply current for each branch, so the total current is the sum of all the branch currents.

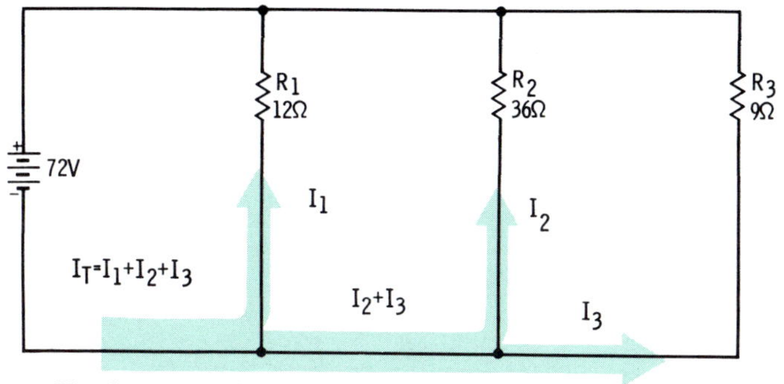


**Total current is the sum of the branch currents.**

**Q3. What is common to all light circuits in a car?**

**Q4. What sort of a load does a car battery have?**

**Q5. The total current in a parallel circuit is the ___ of all the branch currents.**

**Your Answers Should Be:**

**A3. The voltage source.**

A4. Since more than one circuit can be switched into use, the battery can have many different loads. The total load is the sum of the individual loads.

A5. The total current in a parallel circuit is the **sum** of all the branch currents.

### Calculating Current in a Parallel Circuit

To find the current in each branch of a parallel circuit you must apply Ohm's law; that is, E/R equals I for each branch. In the diagram on the preceding page, $I_1$ (branch 1) will equal 72/12, $I_2$ (branch 2) will equal 72/36, and $I_3$ (branch 3) will equal 72/9. The current in branch 1, therefore, is 6 amps; in branch 2 there is 2 amps; and in branch 3 there is 8 amps. Total current supplied by the source is 16 amps.

## CALCULATING TOTAL RESISTANCE

Since $I_T$ in a parallel circuit is equal to the sum of all the branch currents, $R_T$ is equal to $E/I_T$. Also, $I_T$ equals the source voltage divided by $R_T$, and I (any branch) equals the source voltage divided by the resistance of that branch.

$$I_1 = \frac{E}{R_1},\ I_2 = \frac{E}{R_2},\ \text{and}\ I_3 = \frac{E}{R_3}$$

Substituting for $I_T$, $I_1$ (branch 1), etc.

$$(1)\quad I_T = I_1 + I_2 + I_3$$

$$(2)\quad \frac{E}{R_T} = \frac{E}{R_1} + \frac{E}{R_2} + \frac{E}{R_3}$$

E can be any value you choose for the solution of $R_T$. This leads to two possible methods of solving for $R_T$.

**First Process**—$R_T$ can be found by assuming any value for E that is easy to work with. Then, divide each resistance into E to find I for each branch.

As an example, imagine a parallel circuit having resistances of 100, 200, and 200 ohms. Select a value for E that permits an easy solution for all branches, such as 100 volts. Therefore, $E/R_1 = I_1$, $E/R_2 = I_2$, and $E/R_3 = I_3$. $I_1 = 100/100 = 1$ amp, $I_2 = 100/200 = 0.5$ amp, and $I_3 = 100/200 = 0.5$ amp. Now, find the sum of the branch currents to determine $I_T$.

$$I_T = I_1 + I_2 + I_3 = 1a + 0.5a + 0.5a = 2a$$

You now know $I_T$ for a selected E. All that remains in order to find $R_T$ is to divide the assumed voltage by the amount of total current the parallel resistances allow to flow.

$$R_T = \frac{E}{I} = \frac{100}{2} = 50 \text{ ohms}$$

The shorthand statement that describes all of the operations used in finding the total resistance (50 ohms) is:

$$R_T = \frac{E}{I}$$

Since,

$$I_T = \frac{E}{R_1} + \frac{E}{R_2} + \frac{E}{R_3}$$

and,

$$E = 100 \text{ volts}$$

then,

$$R_T = \frac{100}{1+0.5+0.5} = \frac{100}{2} = 50 \text{ ohms}$$

**Second Process**—Looking at the expression $I_T = I_1 + I_2 + I_3$ and using its equivalent expression, $E/R_T = E/R_1 + E/R_2 + E/R_3$, you can arrive at the same expression for $R_T$ as before.

$$\frac{E}{R_T} = \frac{E}{R_1} + \frac{E}{R_2} + \frac{E}{R_3}$$

Now, follow these steps closely.

1. $\left(\frac{E}{R_T}\right)R_T = \left(\frac{E}{R_1} + \frac{E}{R_2} + \frac{E}{R_3}\right)R_T$ (Multiplying both sides by $R_T$.)

2. $E\left(\frac{R_T}{R_T}\right) = \left(\frac{E}{R_1} + \frac{E}{R_2} + \frac{E}{R_3}\right)R_T$ ($R_T$'s on left cancel.)

(Divide both sides by $I_T$ and cancel the like quantities on right.)

$$3.\quad \frac{E}{\frac{E}{R_1}+\frac{E}{R_2}+\frac{E}{R_3}} = \frac{\left(\frac{E}{R_1}+\frac{E}{R_2}+\frac{E}{R_3}\right)R_T}{\frac{E}{R_1}+\frac{E}{R_2}+\frac{E}{R_3}}$$

$$4.\quad \frac{E}{\frac{E}{R_1}+\frac{E}{R_2}+\frac{E}{R_3}} = R_T$$

$R_1 = 100\ \Omega$, $R_2 = 200\ \Omega$, $R_3 = 200\ \Omega$, and 100 volts was previously selected for E. Now, substitute and perform the arithmetic to solve for $R_T$.

$$R_T = \frac{100V}{\frac{100V}{100\Omega}+\frac{100V}{200\Omega}+\frac{100V}{200\Omega}}$$

$$= \frac{100V}{1\text{ amp} + 0.5\text{ amp} + 0.5\text{ amp}}$$

$$= \frac{100V}{2\text{ amp}} = 50\text{ ohms}$$

The two processes have led you to the common expression used in the field for the total resistance of a parallel circuit. ("1" is used as the common value for E.)

$$R_T = \frac{1}{\frac{1}{R_1}+\frac{1}{R_2}+\frac{1}{R_3}+\cdots+\frac{1}{R_n}}$$

This is called a **reciprocal** expression because there are operations in which a value is divided into one. You will find this true for many different mathematical operations. To improve your ability to apply them, examine such expressions to determine the several arithmetic operations each contains. The development of mathematical processes is not often explained in technical literature. Difficult expressions, such as the reciprocal formula above, contain simple arithmetic steps that make the "shorthand" expression reasonable and meaningful. Find and analyze these steps. Acceptance or memorization is of little value unless you understand the reasoning behind each expression.

You should now review the entire process (the solution of $R_T$) until all steps are clear and appear reasonable to you. Evaluate each new step with questions, such as, "Why perform this operation?" or "What arithmetic operations are the symbols representing?" Remember that any operation you will ever follow is nothing more than arithmetic applied with logical rules and sequences.

Many times you will be faced with a "new" mathematical expression (at least it will be strange to you). If you determine the answer to the two questions above, you will find there is a simple hidden reason or step.

The question below asks you to find the solution for $R_T$, using either process. If you do not use the reciprocal form, E can have any value you choose to select. Your selection should be one that permits simple number operations with arithmetic. A hint for the selection of E—always select a value for E equal to or larger than the largest resistance. Another hint—try to select a value for E which results in a whole number when divided by any value of R. Don't forget the Ohm's law relationship where $R_T = E/I_T$.

**Q6. What is the total resistance of the following circuits?**

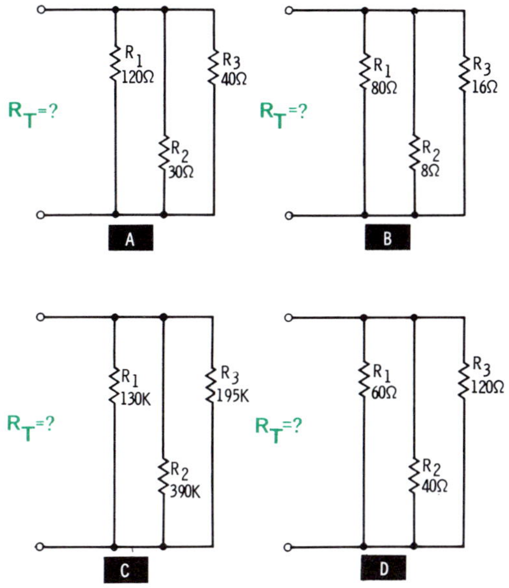

**Your Answers Should Be:**

**A6.** (a) Select E of 120V

$$R_T = \frac{E}{I_T} = \frac{120V}{\frac{120V}{120\Omega} + \frac{120V}{30\Omega} + \frac{120V}{40\Omega}}$$

$$R_T = \frac{120V}{1a + 4a + 3a} = \frac{120V}{8a} = \textbf{15 ohms}$$

(b) Select E of 80V

$$R_T = \frac{E}{I_T} = \frac{80V}{\frac{80V}{80} + \frac{80V}{8} + \frac{80V}{16}}$$

$$R_T = \frac{80V}{1a + 10a + 5a} = \frac{80V}{16a} = \textbf{5 ohms}$$

(c) Select an E of 390V

$$R_T = \frac{E}{I_T} = \frac{390V}{\frac{390V}{130K} + \frac{390V}{390K} + \frac{390V}{195K}}$$

$$R_T = \frac{390V}{3ma + 1ma + 2ma} = \frac{390V}{6ma} = \textbf{65K}$$

(d) Select an E of 120V

$$R_T = \frac{E}{I_T} = \frac{120V}{\frac{120V}{60} + \frac{120V}{40} + \frac{120V}{120}}$$

$$R_T = \frac{120V}{2a + 3a + 1a} = \frac{120V}{6a} = \textbf{20 ohms}$$

### Total R in a Two-Branch Circuit

Another method for finding the total resistance of a two-branch parallel circuit is the **product-over-the-sum** process. Given the values of two resistors in parallel, multiply one times the other and divide by the sum of the two.

$$R_T = \frac{R_1 \times R_2}{R_1 + R_2}$$

The total resistance of the parallel circuit below is 24 ohms. This can be determined by using the product-over-the-sum process.

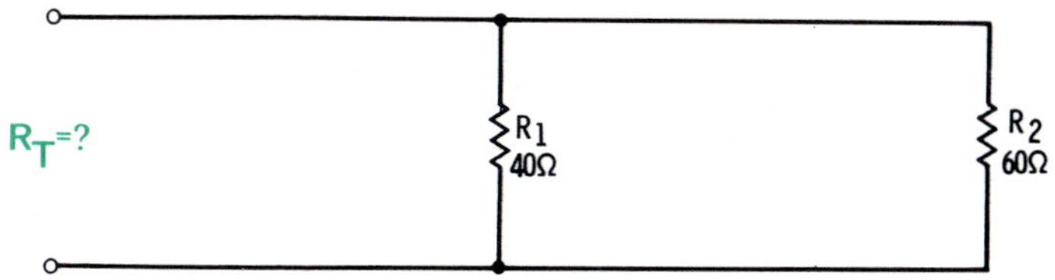


$$R_T = \frac{40 \times 60}{40 + 60} = \frac{2{,}400}{100} = 24 \text{ ohms}$$

$$R_T = \frac{E}{\dfrac{E}{R_1} + \dfrac{E}{R_2}} = \frac{E}{\dfrac{E_{R2} + E_{R1}}{R_1 \times R_2}} = \frac{E(R_1 \times R_2)}{E(R_2 + R_1)}$$

To find the total resistance of a parallel circuit containing resistors of equal value in parallel, all that needs to be done is to divide the value of one of the resistors by the number of resistors in parallel.

If a parallel circuit, for example, contains three 90-ohm resistors, $R_T$ can be determined by dividing 90 by 3.

$$R_T = \frac{90}{\dfrac{90}{90} + \dfrac{90}{90} + \dfrac{90}{90}} \quad \text{(selecting an E of 90)}$$

$$= \frac{90}{1 + 1 + 1}$$

$$= \frac{90}{3} = 30 \text{ ohms}$$

**Q7. What is the total resistance of these circuits?**

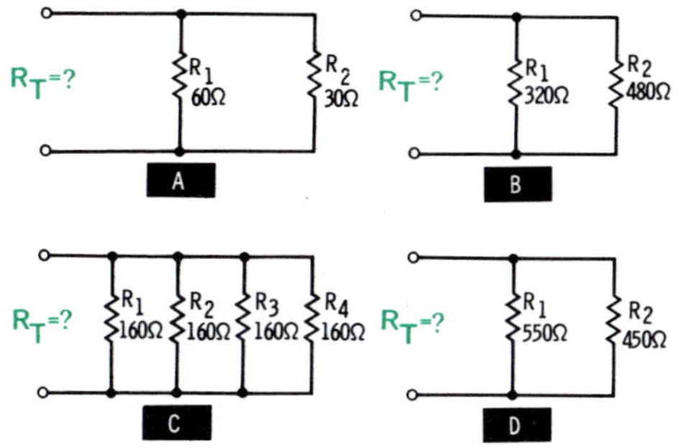

**Your Answers Should Be:**

A7. (a) $R_T = \frac{R_1 \times R_2}{R_1 + R_2} = \frac{60 \times 30}{60 + 30} = \frac{1,800}{90} = $ **20 ohms**

(b) $R_T = \frac{320 \times 480}{320 + 480} = \frac{153,600}{800} = $ **192 ohms**

(c) $R_T = \frac{160}{4} = $ **40 ohms**

(d) $R_T = \frac{550 \times 450}{550 + 450} = \frac{247,500}{1,000} = $ **247.5 ohms**

### Equivalent Resistance

Just as in the series circuit, the resistances in a parallel circuit can be represented by an equivalent. The equivalent indicates the load ($R_T$) which the source must work into. Since $I_T$ is the easiest unknown to find (it is the sum of all the currents), it becomes a simple task to divide $I_T$ into the source voltage to find $R_T$.

By the same reasoning, it is easy to find the power that the source must supply.

$$P = IE$$

In this case:

$$P_T = (I_1 \times E) + (I_2 \times E) + (I_3 \times E) = I_TE$$

## TYPICAL APPLICATIONS

A parallel circuit is used where current is to be divided. This is similar to the action of the series circuit, except that voltage was divided.

**Current Meter**—An ammeter is used to indicate the amount of current flowing in a circuit. Very often the meter used will be one that has a full-scale deflection, indicating an amount of current much less than the circuit current to be measured. There is a method of bypassing the meter with the excess current. This is called **shunting** the meter. With a shunt, the meter reads a percentage of the total current. This means normal current flows through the entire parallel network (consisting of the meter and shunt), with only a small portion flowing through the meter.

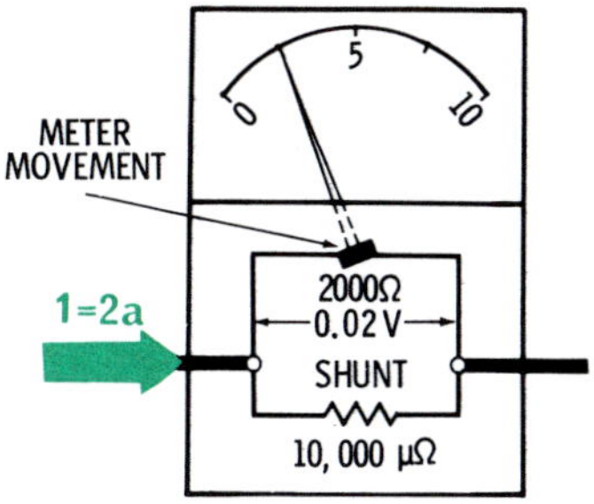


**An ammeter with a shunt.**

This meter has a movement with 2,000 ohms of resistance and is shunted by a 10,000 micro-ohm (0.01 ohm) resistor. This permits 10 $\mu$a of current to flow through the movement and 2 amps (minus the 10 $\mu$a) of current to flow through the shunt. These may not be the exact values for a particular multimeter, but all multimeters employ similar ratios.

**Q8. What is the total current in the following circuits?**

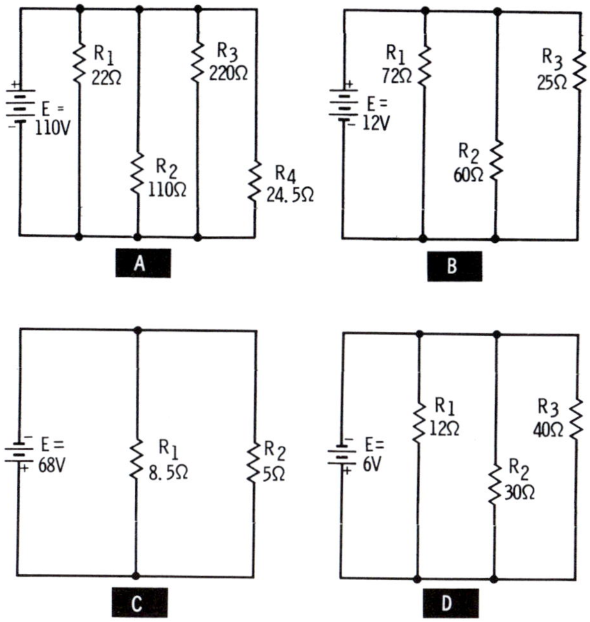


**Q9. What is the power dissipated by each resistor and the total power for each circuit above?**

**Your Answers Should Be:**

**A8.** (a) $I_T = I_1 + I_2 + I_3 + I_4 = 110/22 + 110/110 + 110/220 + 110/24.5 = 5$ amps $+ 1$ amp $+ 0.5$ amp $+ 4.489$ amps $=$ **10.989 amps**

(b) $I_T = 12/72 + 12/60 + 12/25 = 0.167$ amp $+ 0.2$ amp $+ 0.48$ amp $=$ **0.847 amp**

(c) $I_T = 68/8.5 + 68/5 = 8$ amps $+ 13.6$ amps $=$ **21.6 amps**

(d) $I_T = 6/12 + 6/30 + 6/40 =$ **0.85 amp**

**A9.** (a) $P_1 = 5 \times 110 =$ **550 watts;** $P_2 = 1 \times 110 =$ **110 watts;** $P_3 = 0.5 \times 110 =$ **55 watts;** $P_4 = 4.489 \times 110 =$ **493.79 watts;** $P_T = 10.989 \times 110 =$ **1,208.79 watts**

(b) $P_1 = 0.167 \times 12 =$ **2.004 watts;** $P_2 = 0.2 \times 12 =$ **2.4 watts;** $P_3 = 0.48 \times 12 =$ **5.76 watts;** $P_T = 0.847 \times 12 =$ **10.164 watts**

(c) $P_1 = 8 \times 68 =$ **544 watts;** $P_2 = 13.6 \times 68 =$ **924.8 watts;** $P_T = 216 \times 68 =$ **1,468.8 watts**

(d) $P_1 = 0.5 \times 6 =$ **3 watts;** $P_2 = 0.2 \times 6 =$ **1.2 watts;** $P_3 = 0.15 \times 6 =$ **0.9 watt;** $P_T = 0.085 \times 6 =$ **5.1 watts.**

**Switches in Parallel**—Switches are connected in series in some cases and in parallel in others.

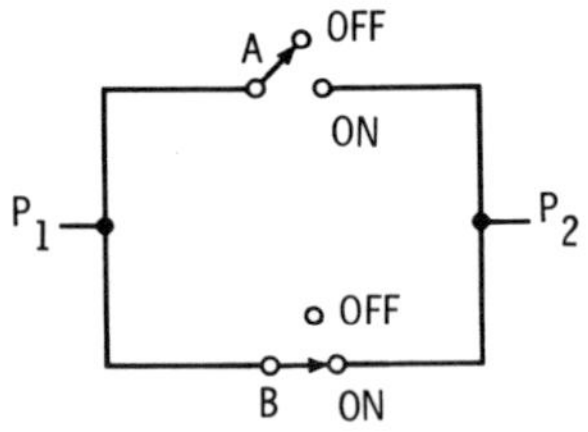


**Separately actuated switches in parallel.**

As can be seen in the figure above, the switches can form a closed-circuit condition from $P_1$ to $P_2$ if either one or the other or both are closed. To obtain an open circuit from $P_1$ to $P_2$, both switches must be open. Another form of the parallel switch is the type which has two or more poles actuated by the same mechanical element.

Group actuated switches in parallel.

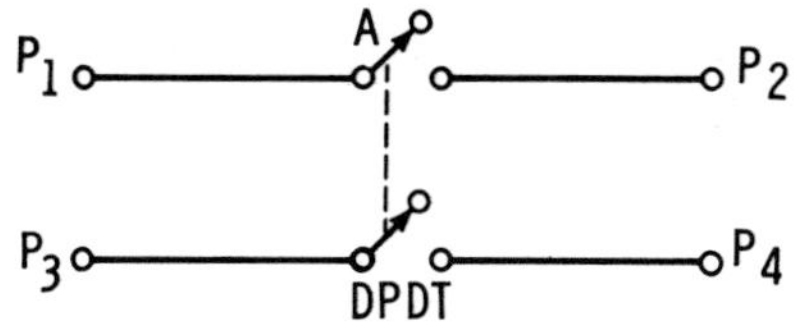


The dashed line indicates that the two poles of the switch are actuated at the same time.

The rotary action available in the application of the wafer switch permits many parallel operations. In the illustration

Wafer switch parallel operation.

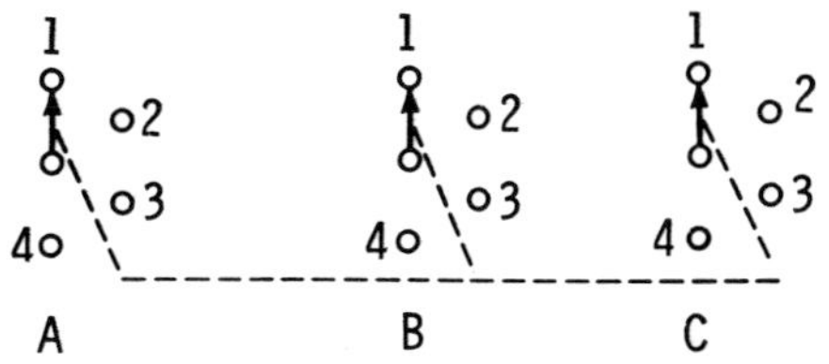


above, the switch has three wafers (A, B, and C). Again, the dashed line represents mechanical connection whereby all wafers are actuated at the same time. In this case the switch is said to be a **three-** (triple) **pole, quadruple-throw** (four position), or **TPQT** (triple-pole, quadruple-throw).

**Batteries in Parallel**—For heavy-duty operation, batteries of equal voltage are constructed with plates in parallel or connected with many cells in parallel. This means they require more charging current to become fully charged. Yet, in another sense, more current must be drawn to cause the batteries to be discharged.

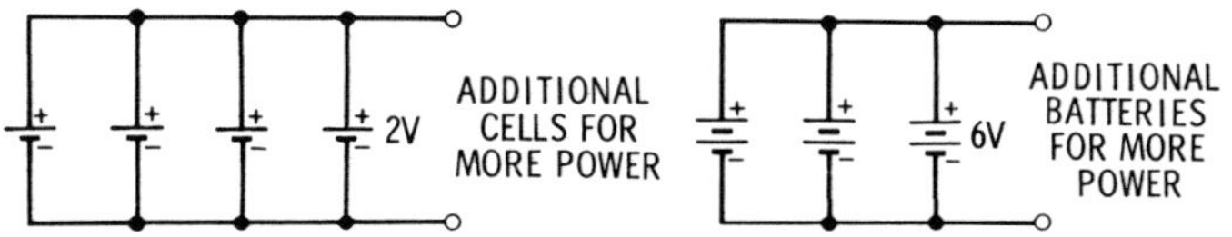


Batteries in parallel.

**Q10. What precautions must you take before connecting batteries in parallel?**

**Q11. Could you use a lamp bulb for a shunt, or parallel, resistance in a circuit? How would you determine the resistance of the lamp?**

**Your Answers Should Be:**

**A10.** You must make sure all batteries to be connected in parallel have **the same voltage.**

**A11.** **Yes,** you could use a lamp for a shunt in a circuit. The wattage and voltage ratings permit you to calculate the resistance. That is , $\mathbf{I} = \frac{\mathbf{P}}{\mathbf{E}}$ and then $\mathbf{R} = \frac{\mathbf{E}}{\mathbf{I}}$.

## WHAT YOU HAVE LEARNED

1. A parallel circuit is a combination of two or more basic circuits connected to a common voltage source.
2. Batteries may be connected in parallel to produce power for a longer time.
3. Switches may be connected in parallel to form parallel turn-on operations; yet they all have to be turned off to open the circuit.
4. Filaments in vacuum tubes may be connected across a common voltage source.
5. Each branch of a parallel circuit may have a different current flowing through it. All branches will have the same voltage applied.
6. To find the total current in a parallel circuit, you determine the sum of all the branch currents.

$$I_T = I_1 + I_2 + I_3 + \ldots$$

7 To find the total resistance of a parallel circuit, divide the applied voltage by the total current in the circuit.

$$R_T = \frac{E}{I_T}$$

8. If the resistance is the only factor known, you may use any value for E that permits simple arithmetic operations for determining the current in each branch. After the total current has been computed, the same value for E must be used to determine total resistance.

# 5

# Combined Series and Parallel Circuits

## *What You Will Learn*

**This chapter contains applications of the series and parallel fundamentals you learned in preceding chapters. You will now learn how to determine the direction of current flow in series-parallel circuits. When you complete this chapter, you will be able to reduce combinations of series and parallel circuits to a series equivalent. Also you will be able to determine and compute currents and voltages in each part of the circuit, and to apply Kirchhoff's law properly when needed. A good understanding of the material in this chapter prepares you for complex electrical and electronic circuit examination, using a simple step-by-step logical process.**

### IDENTIFYING INDIVIDUAL CIRCUITS

A basic electrical circuit consists of a source, a load, and conductors that connect the source to the load. The basic circuit was discussed in preceding chapters as if it were a loop. An applied voltage will cause a current to flow through a resistive element in a complete round-trip path. There are two ways to explain the direction of current flow—conventional and electron. If you employ the conventional theory for current flow, you describe all electrical flow in terms of positive ions in motion. Electron flow (the concept used in this text) states that current is the movement of electrons.

According to this theory, electrons leave the negative terminal of a source, move through the circuit, and return to the positive terminal.

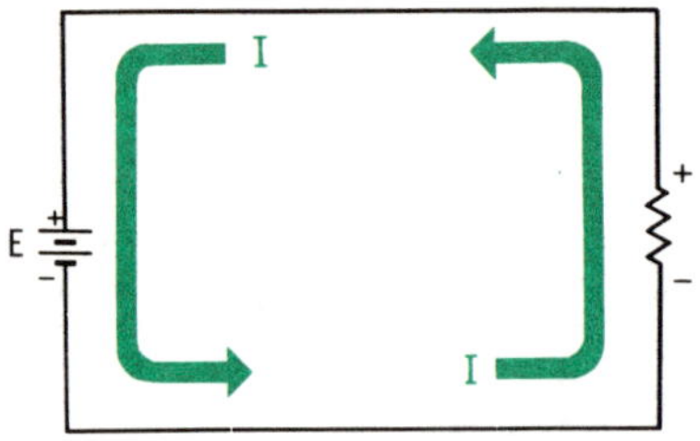


**Electronic current flow.**

## SERIES CIRCUITS

A series circuit is a basic circuit with all electrical components connected end to end. The key for determining the total voltage when there are two or more sources having different voltages and polarities is to find the potential difference between them. Next, assign the polarity of the larger voltage to the output terminals.

In a series circuit the total resistance is calculated by finding the sum of the individual resistances. Current in a series circuit is found by dividing the source voltage by the total resistance.

Since the total current ($I_T$) is flowing through all resistances, the voltage drop across a series resistance can be found by multiplying the current by the individual value of resistance. In other words, $E = IR$. The voltage drop across any resistance is equal to the current through the resistance multiplied by the value of the resistance.

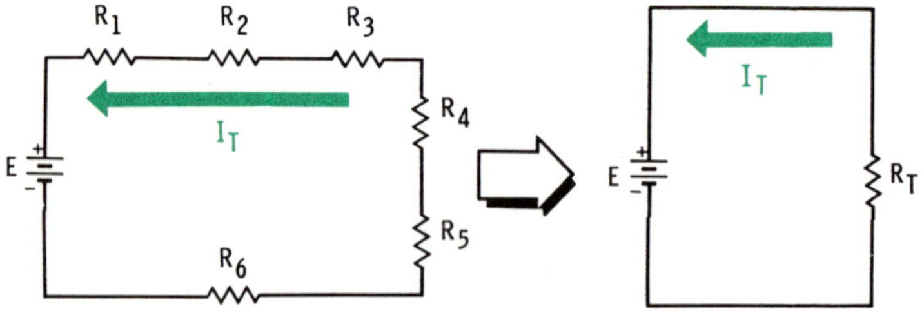


$$R_T = R_1 + R_2 + R_3 + R_4 + R_5 + R_6$$

$$I_T = \frac{E}{R_T}$$

A series circuit can have more than one source connected in such a manner as to aid or to oppose one another.

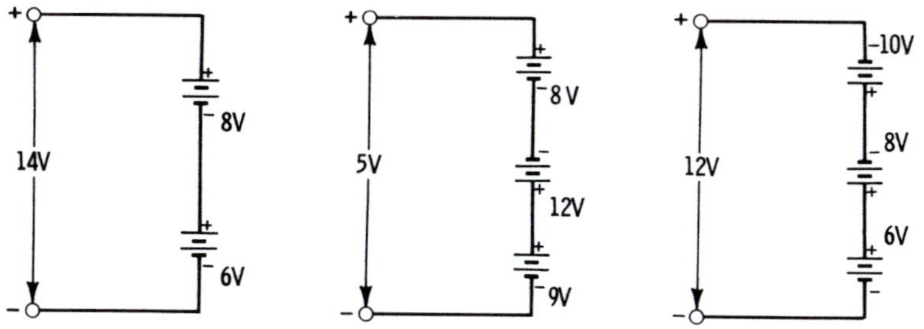


AIDING OPPOSING OPPOSING

**Q1. In which direction does current flow inside the source when the conventional current-flow theory is being employed?**

**Q2. Draw a basic circuit and indicate the direction of current flow. (Use the electron theory.)**

**Q3. Can a basic circuit be considered a series circuit?**

**Q4. What is the total voltage at the terminals of the following sources?**

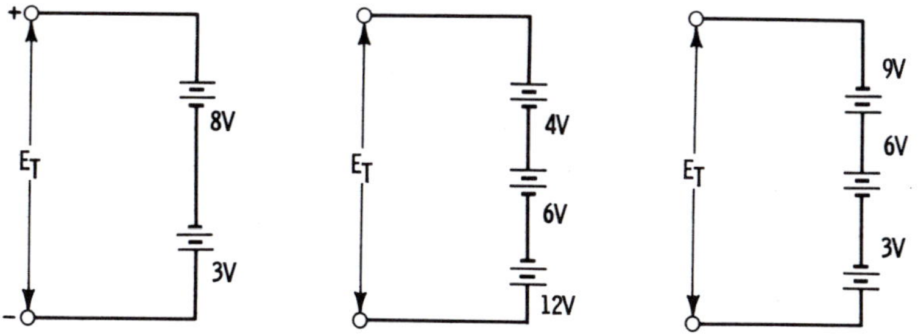


A B C

**Q5. How does $I_T$ compare with the value of current through any one of the resistances in the following circuit?**

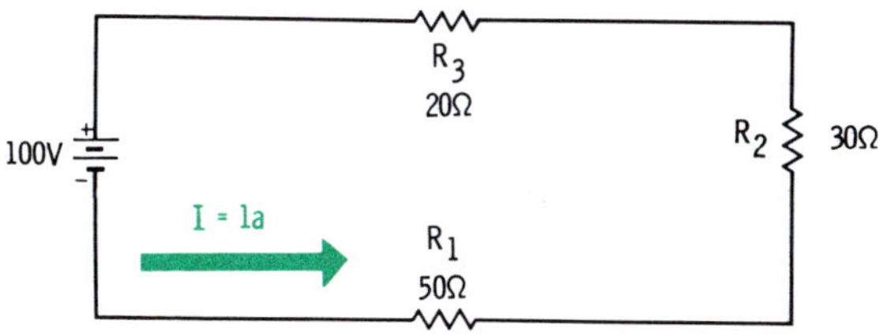

**Your Answers Should Be:**

**A1.** When conventional current theory is being employed, current will flow **from the negative terminal to the positive terminal** within the source.

**A2.**

**A3.** A basic circuit **is** a fundamental series circuit.

**A4.** A. Negative to positive, bottom to top, **5 volts.**

B. Negative to positive, top to bottom, **2 volts.**

C. Negative to positive, top to bottom, **6 volts.**

**A5.** $I_T$ has **the same value** as the current through any of the resistances.

## Voltage Division

When current flows through a resistance, a voltage can be measured across it. The voltage across each resistor in a series circuit is equal to the total current times the value

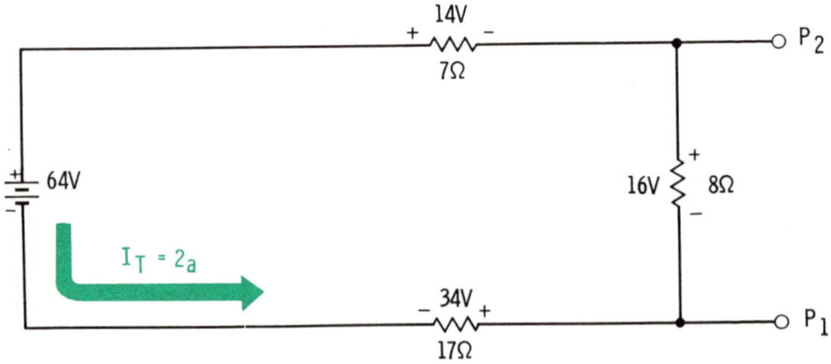


of its resistance. That is, the 17-ohm resistor in the above circuit will have 34V (2 amps times 17 Ω) developed across it. The other voltages are determined similarly. Thus the voltage between $P_1$ and $P_2$ is 16V.

## IR Drop

What is another way of expressing voltage drop across a resistor? Since voltage equals IR, it may be called **IR drop.** This is the same as saying the voltage developed across $R_1$, $R_2$, etc. The resistor voltage drops would be $IR_1$, $IR_2$, etc.

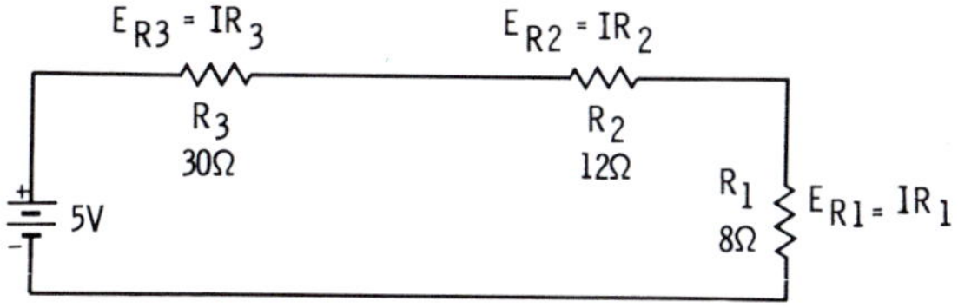


$$I_T = \frac{E}{R_1 + R_2 + R_3} = \frac{5}{50} = 0.1 \text{ amp}$$

The voltage across $R_1$ equals $IR_1$, or $0.1 \times 8 = 0.8V$. The voltage across $R_2$ equals $IR_2 = 0.1 \times 12 = 1.2V$. The voltage across $R_3$ equals $IR_3 = 0.1 \times 30 = 3V$. The total IR drop is equal to the source voltage.

$$0.8V + 1.2V + 3.0V = 5V$$

**Q6. What is the voltage drop across the $R_1$ resistors in the following circuits?**

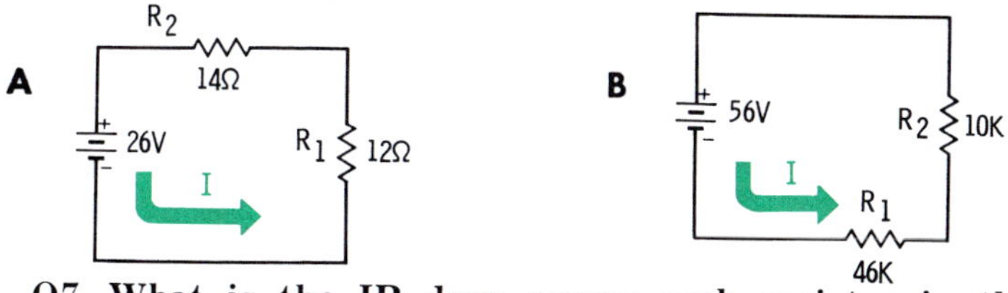


**Q7. What is the IR drop across each resistor in the following circuits?**

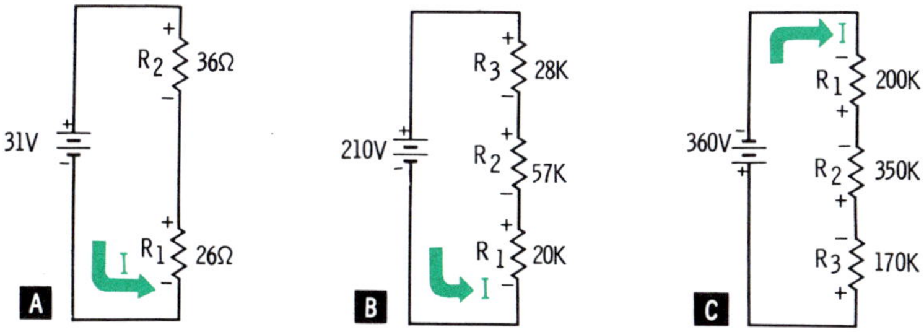


**Q8. How does the voltage drop across the 36-ohm resistor affect the amount of voltage drop across the 26-ohm resistor in (a) above?**

**Your Answers Should Be:**

**A6.** (a) **12 volts** = 1 amp times 12 Ω.

(b) **46 volts** = 1 ma times 46K.

**A7.** (a) $IR_1$ = **13V**, and $IR_2$ = **18V** ($I_T$ = 0.5 amp)

(b) $IR_1$ = **40V**, $IR_2$ = **114V**, and $IR_3$ = **56V** ($I_T$ = 2 ma)

(c) $IR_1$ = **100V**, $IR_2$ = **175V**, and $IR_3$ = **85V** ($I_T$ = 0.5 ma)

**A8.** The voltage across the 26-ohm resistor will be **decreased** from the source voltage by an amount equal to the voltage drop across the 36-ohm resistor.

## PARALLEL CIRCUITS

A parallel circuit has two or more loads (resistances) connected across a source. The current flow through each resistance depends on the amount of that resistance. The total load ($R_T$) can be determined by dividing the source voltage by the total current. The total current equals the sum of the separate branch currents. $I_T = I_1$ (branch 1) $+ I_2$ (branch 2) + etc.

How does this compare to the series circuit? In a series circuit, the same current flows through all resistances, and thus a portion of the source voltage is dropped (proportional to the value of resistance) across each resistance.

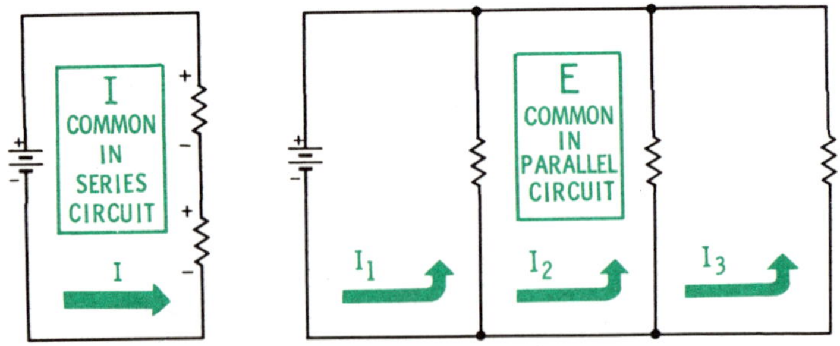


A parallel circuit has a common supply voltage, and thus a possible different current through each resistance (branch), depending on the value of the resistance.

These two factors must be remembered in all operations concerned with calculations in both series and parallel circuits—to work problems in series circuits, employ the common current as the working component of Ohm's law along with the different resistances; to work problems in parallel circuits, employ the common voltage as the working component of Ohm's law. Let a parallel circuit with two resistors serve as an example.

$$(1)\quad R_T = \frac{E}{\frac{E}{R_1} + \frac{E}{R_2}} = \frac{E}{I_T}$$

This can also be stated as:

$$(2)\quad R_T = \frac{R_1 \times R_2}{R_1 + R_2}$$

If all resistors are equal:

$$(3)\quad R_T = \frac{\text{R (value of one resistor)}}{\text{the number of resistors in parallel}}$$

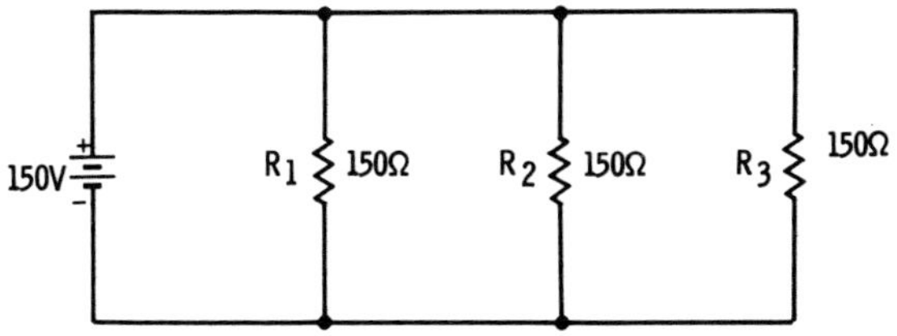


Using (1):

$$R_T = \frac{150V}{\frac{150V}{150} + \frac{150V}{150} + \frac{150V}{150}} = \frac{150V}{1a + 1a + 1a}$$

$$R_T = \frac{150V}{3a} = 50 \text{ ohms}$$

However, using (3):

$$R_T = \frac{150}{3} = 50 \text{ ohms}$$

**Q9. When working problems from a schematic, what value of resistance do the conductors have?**

**Your Answer Should Be:**

**A9.** The conductors are considered to have a **zero resistance,** unless otherwise stated, when working problems from a schematic.

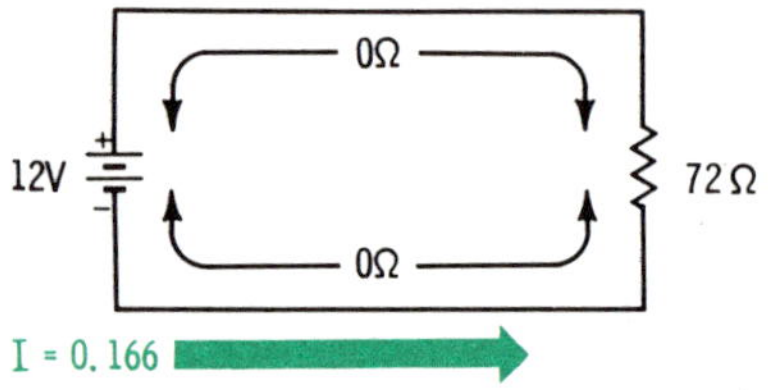


## SERIES AND PARALLEL COMBINATIONS

From the basic series and parallel circuits there are many combinations possible which will contain both series and parallel characteristics. The diagram below demonstrates the two in combination.

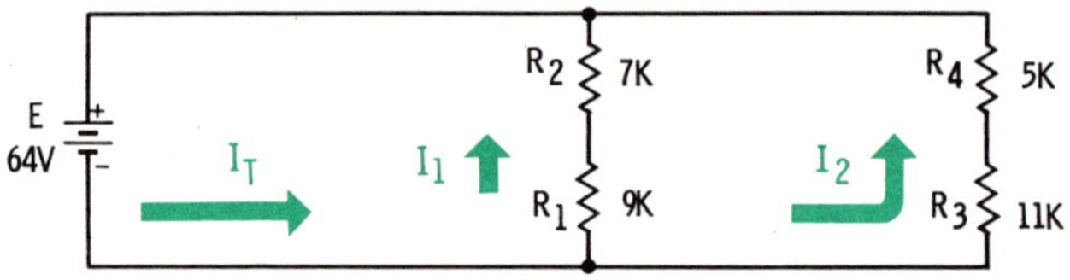


The first parallel branch consists of two resistors in series ($R_1$ and $R_2$). The second parallel branch contains two more resistors in series ($R_3$ and $R_4$). Many solutions can be derived from the problems which may stem from such arrangements. For instance, the first branch resistance is determined by finding $R_T$ for the series circuit. The second branch resistance is determined by the same process. The total resistance can be determined by employing the $R_T = E/I_T$ expression. In this case, $I_T$ would be the result of the sum of $I_1$ in branch 1 plus $I_2$ in branch 2. This is the parallel circuit process.

The possible routes of solution for the circuit above may be evaluated using many processes. The best approach is always the direct application of Ohm's law expressions. In this case, $R_T = E/I_T$. E is known but $I_T$ must be determined. $I_T = I_1 + I_2$. $I_1 = E/R_T$ for branch 1, and $I_2 = E/R_T$ for branch 2. Since they are series branches, $R_T$ is the sum of the resistances. That is, $R_{T1} = R_1 + R_2$ for branch 1, and $R_{T2} = R_3 + R_4$ for branch 2.

The total resistance of each branch can be used to find the branch currents, the branch currents to find the total circuit current, and the total circuit current to find the total circuit resistance ($R_T$). Other combination circuits may take on almost any form. The forms are limited only by the designer's need to construct a circuit which will perform a specific function.

Another example of a combination circuit is shown below:

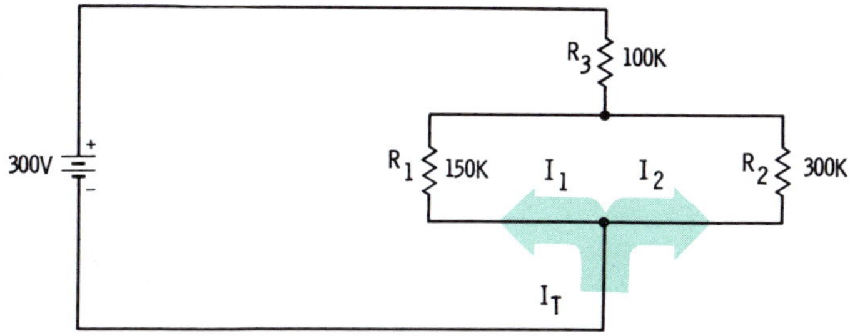


To find the effective load for the 300V source, $R_T$ must be calculated. One approach for finding $R_T$ is to first find the total resistance for the parallel branch ($R_1$ and $R_2$); then the sum of the total parallel resistance plus $R_3$ should be found. After the total resistance has been determined, the total current can be calculated.

The total resistance for the parallel combination of $R_1$ and $R_2$ may be found by employing the product-over-the-sum process. That is:

$$R_{T1} = \frac{R_1 \times R_2}{R_1 + R_2} = \frac{150K \times 300K}{150K + 300K} = 100K$$

for the parallel branch. The final series-circuit resistance becomes $R_T = 100K + 100K = 200K$. The total current for the entire combination is 300V/200K, or 1.5 ma.

**Q10. What is the total resistance in the circuit on the opposite page?**

**Q11. What is the total current in the circuit on the opposite page?**

**Q12. What is the voltage drop across $R_4$?**

**Q13. Draw the equivalent circuit.**

**Your Answers Should Be:**

**A10.** $R_{T1} = R_1$ plus $R_2$ for branch 1. $R_{T2} = R_3$ plus $R_4$ for branch 2.

$R_{T1} = 9K + 7K = 16K$ for branch 1.

$R_{T2} = 11K + 5K = 16K$ for branch 2.

**Total resistance equals 16K/2 = 8K.**

**A11.** $I_T = E/R_T = 64V/8K = 8$ **ma.**

**A12.** The voltage drop across $R_4$ equals $IR_4 = 4 \text{ ma} \times 5K =$ **20V.**

**A13.**

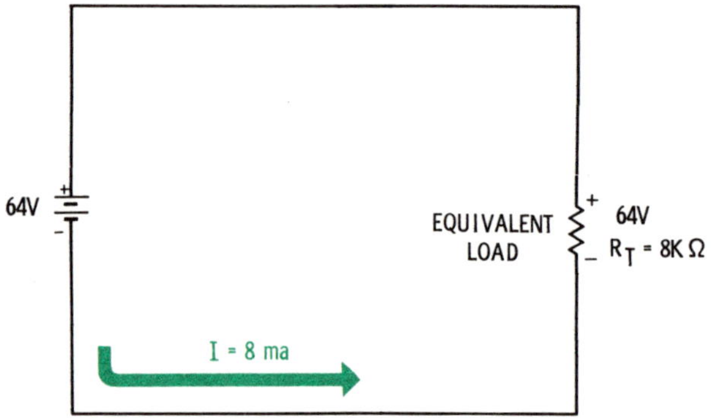


The voltage drop across $R_3$ (diagram on preceding page) is equal to $IR_3$. $IR_3 = 1.5 \text{ ma} \times 100K = 150$ volts. The voltage common to both $R_1$ and $R_2$ is the source voltage minus the voltage drop across $R_3$. This results in a voltage of 150 volts across the two resistors in parallel. $R_1$ will have 1 ma of current flowing through it. $R_2$ will have 0.5 ma of current flowing through it. The total current of the branches (and the circuit) will equal 1.5 ma. The equivalent circuit would consist of a 300-volt source connected to a 200-kilohm load.

The illustration on the next page, under Question 14, shows another example of a combination series-parallel circuit. Included in this illustration is a method recommended for converting this type of circuit to an equivalent series circuit. This conversion simplifies the calculations.

**Q14.** If a 100-volt source is applied to the circuit below, what will be the voltage drop across $R_4$? What will be the current flow through $R_4$?

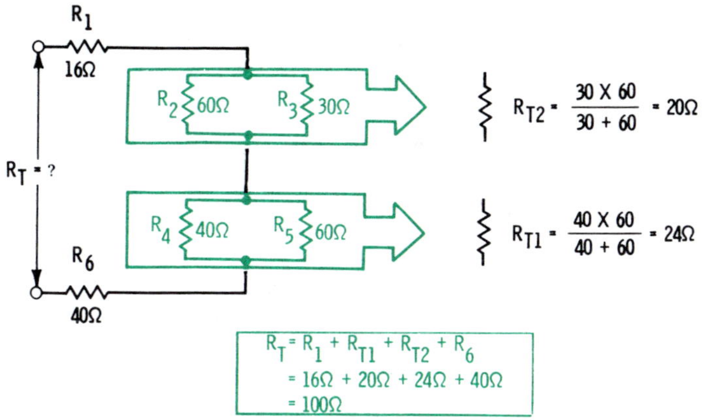


**Q15.** What is the voltage drop across each resistor in the following circuits?

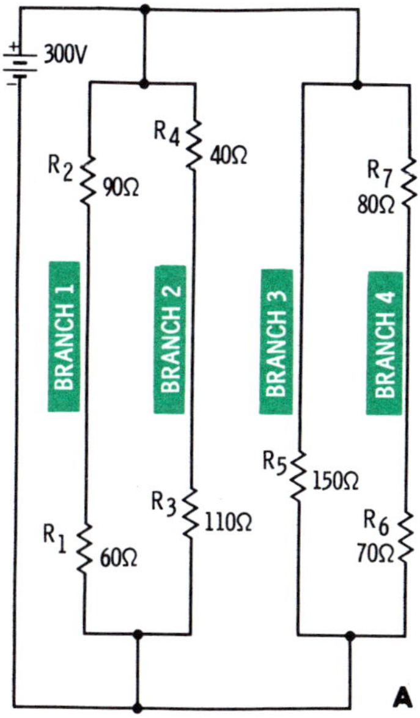


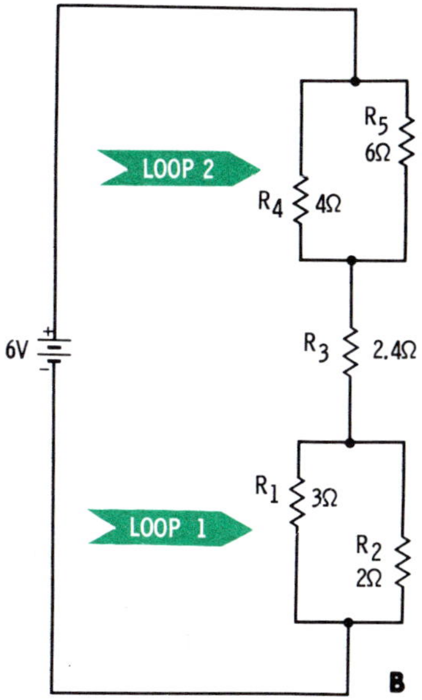

**Your Answers Should Be:**

**A14.** The total current will be **1 ampere.** The current through $R_4$ will be **0.6 amp.** So, $E_{R4} = 24V$.

**A15.** (a) To find the voltage drop across each resistor, you must find the current through each branch. In this case, total current is of no concern.

BRANCH 1: $$I_{loop\ 1} = \frac{300V}{R_1 + R_2} = \frac{300V}{150\Omega} = 2 \text{ amps}$$

$$IR_1 = 2a \times 60\Omega = \mathbf{120V}$$
$$IR_2 = 2a \times 90\Omega = \mathbf{180V}$$

BRANCH 2: $$I_{loop\ 2} = \frac{300V}{R_3 + R_4} = \frac{300V}{150\Omega} = 2 \text{ amps}$$

$$IR_3 = 2a \times 110\Omega = \mathbf{220V}$$
$$IR_4 = 2a \times 40\Omega = \mathbf{80V}$$

BRANCH 3:

$IR_5$ will equal the **supply voltage (300V).**

BRANCH 4: $$I_{loop\ 2} = \frac{300V}{R_6 + R_7} = \frac{300V}{150\Omega} = 2 \text{ amps}$$

$$IR_6 = 2a \times 70\Omega = \mathbf{140V}$$
$$IR_7 = 2a \times 80\Omega = \mathbf{160V}$$

(b) To find the voltage drop across each resistor, proceed as follows:

$$R_{T(loop\ 1)} = \frac{R_1 \times R_2}{R_1 + R_2} = \frac{3 \times 2}{3 + 2} = 1.2\Omega$$
$$R_{T(loop\ 2)} = \frac{R_4 \times R_5}{R_4 + R_5} = \frac{4 \times 6}{4 + 6} = 2.4\Omega$$
$$R_T = R_{T(loop\ 1)} + R_3 + R_{T(loop\ 2)}$$
$$R_T = 1.2\Omega + 2.4\Omega + 2.4\Omega = 6\Omega$$
$$I_T = \frac{E}{R_T} = \frac{6V}{6\Omega} = 1 \text{ amp}$$
$$IR_{(loop\ 1)} = 1a \times 1.2\Omega = \mathbf{1.2V}$$
$$IR_{(loop\ 2)} = 1a \times 2.4\Omega = \mathbf{2.4V}$$
$$IR_3 = 1a \times 2.4\Omega = \mathbf{2.4V}$$
$$IR_T = 1.2V + 2.4V + 2.4V = \mathbf{6V}$$

## KIRCHHOFF'S LAW

Kirchhoff's law defines the distribution of currents and voltages within an electrical circuit. This law is used as a method of checking to see if you have assigned the proper direction for current flow and to see if your arithmetic is correct. It consists of two parts—one for voltages and one for currents. You will find this law a very useful tool when the direction of current is in question and/or when the total voltage or current is to be determined.

Primarily, Kirchhoff's law is a complete circuit application of Ohm's law. It makes use of Ohm's law many times in some circuits. Because of this application and the unique methods employed in handling points in the circuit—one with respect to another—you should master every process.

### Voltage Applications

Upon determining voltage and polarity for each source and across each resistance, choose a point in the circuit and assign a direction of current flow. Find the algebraic sum of all the IR drops plus the source, from the chosen point, around the entire circuit and return. It should be zero.

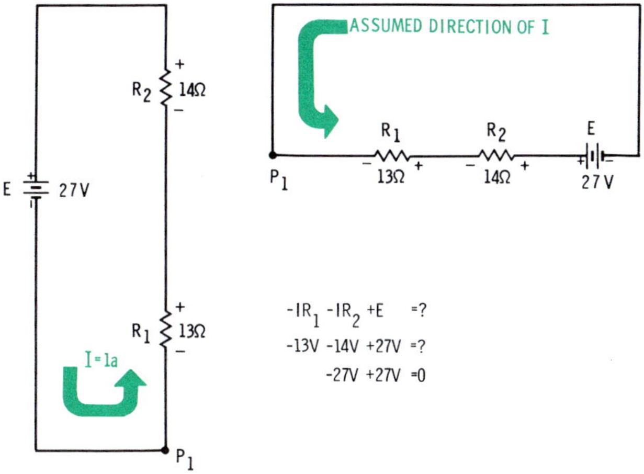


**Kirchhoff's voltage law.**

**The sum of voltages in the loop from point 1 back to point 1 is zero.** The current was assumed to be in the correct direction, and $I_T$ and the IR drops were calculated properly.

Another look at the same circuit, with an opposite direction of current assignment, is as follows.

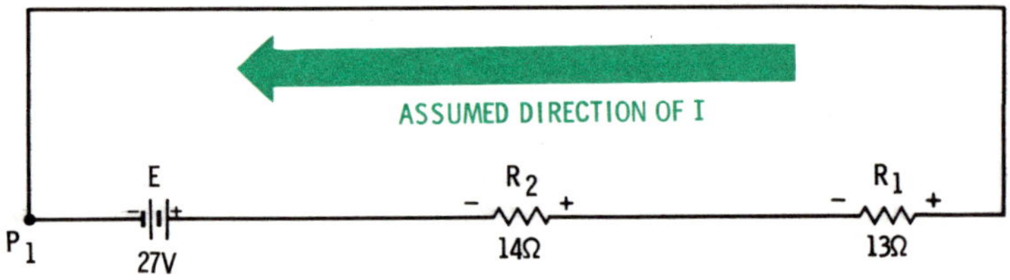


$$E+IR_2+IR_1=?$$
$$(-27V)+(-14V)+(-13V)=-54V$$

The sum of the voltages in the loop from point 1 back to point 1 equals —54V. Therefore the current was assumed to be going in the wrong direction.

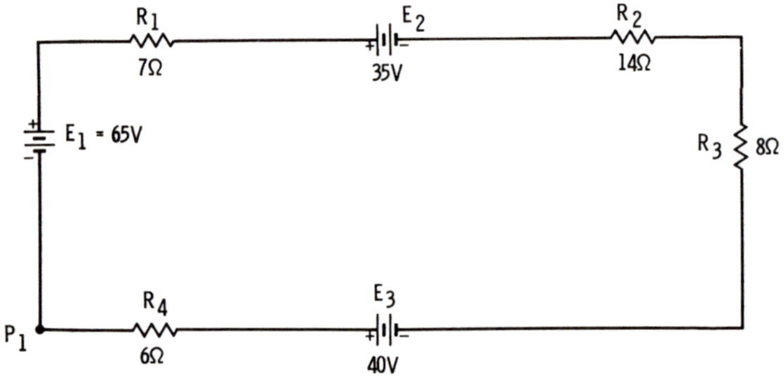


(1) $E_T = +E_3 - E_2 + E_1 = (+40V)+(-35V)+(+65V)=+70V$

(2) $R_T = 6\Omega + 8\Omega + 14\Omega + 7\Omega = 35\Omega$

(3) $I = \frac{E_T}{R_T} = \frac{70V}{35\Omega} = 2a$

(4) $(-IR_4)+(+E_3)+(-IR_3)+(-IR_2)+(-E_2)+(-IR_1)+(+E_1)=?$

Calculating the IR drop across each resistor: $IR_4 = 12V$, $IR_3 = 16V$, $IR_2 = 28V$, and $IR_1 = 14V$. Kirchhoff's application (4) above: —12V + 40V — 16V — 28V — 35V — 14V + 65V = ? 105V — 105V = 0V.

## Current Application

Finding the sum of the voltages around the circuit for the series network is one basic application of Kirchhoff's law. Finding the sum of the currents is another application and is the process employed in a parallel circuit. Kirchhoff's current law states that **the current flow away from a given point in a circuit must equal the current flow to that point.**

**Q16. What is the $E_T$ for the following circuit?**

**Q17. What is the $R_T$ for the following circuit?**

**Q18. What is the I for the following circuit?**

**Q19. Write the complete Kirchhoff expression for the following circuit, starting at $P_1$.**

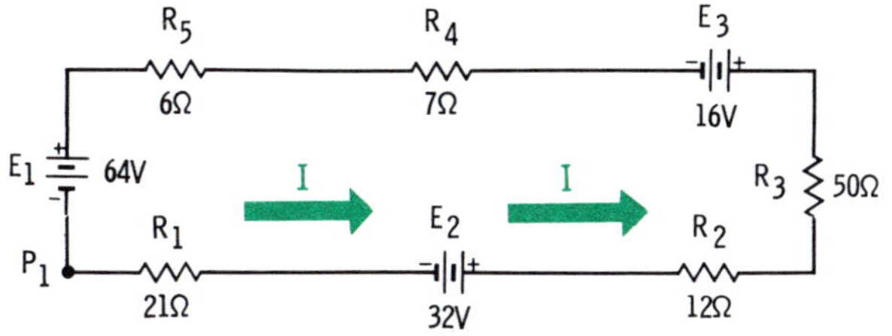


**Q20. Write Kirchhoff's current expression for the following circuit.**

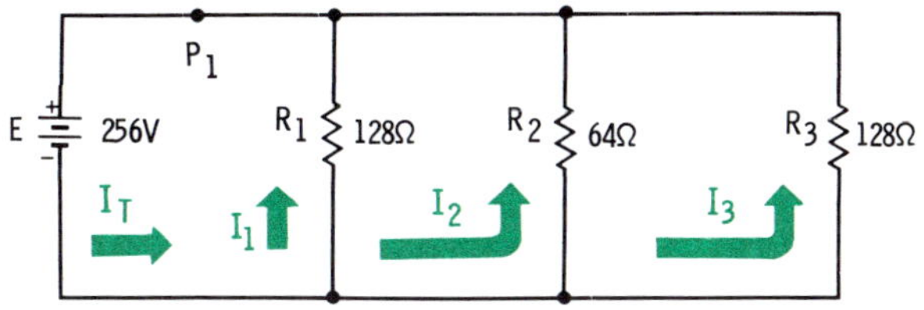


**Q21. What is the voltage at point 1 and the current through R (dropping resistance) for the following circuits?**

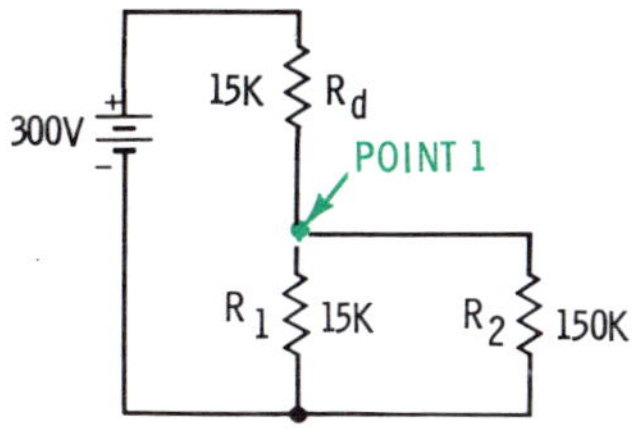


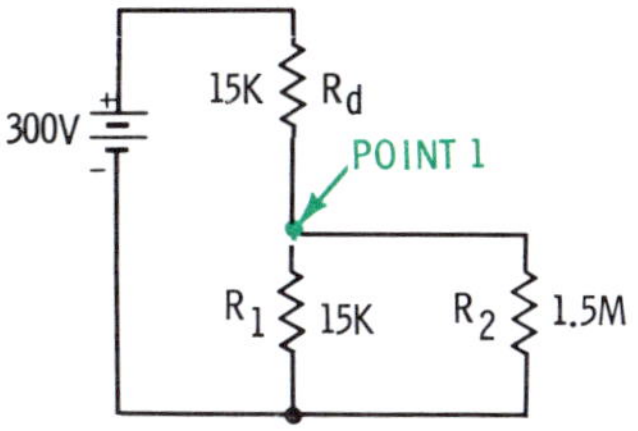

**Your Answers Should Be:**

**A16.** $E_T = 64V - 32V + 16V = \mathbf{48V}$

**A17.** $R_T = R_1 + R_2 + R_3 + R_4 + R_5 = 21\Omega + 12\Omega + 50\Omega + 7\Omega + 6\Omega = \mathbf{96\Omega}$

**A18.** $I = E_T/R_T = 48V/96\Omega = \mathbf{0.5\ amp}$

**A19.** $-IR_1 - E_2 - IR_2 - IR_3 + E_3 - IR_4 - IR_5 + E_1 = -10.5V - 32V - 6V - 25V + 16V - 3.5V - 3V + 64V = 0V$

**A20.** $I_T = I_1 + I_2 + I_3$

$E/R_T = E/R_1 + E/R_2 + E/R_3$

$256V/32\Omega = 256V/128\Omega + 256V/64\Omega + 256V/128\Omega$

**8 amps = 2 amps + 4 amps + 2 amps**

**A21.** $I_T = 300V/R_T$

$$R_T = \frac{R_1 \times R_2}{R_1 + R_2} + R_d$$

$R_T = 28.6K$ for (A) ; $R_T = 29.8K$ for (B)

$\mathbf{I_T = 10.49}$ **ma for (A);** $\mathbf{I_T = 10.06}$ **ma for (B)**

Voltage at $P_1 = 300 - (15K \times 10.49\ ma) = 300 - 157.35 =$ **142.65 volts for (A)**

Voltage at $P_1 = 300 - (15K \times 10.06\ ma) = 300 - 150.9 =$ **149.1 volts for (B)**

## APPLICATION

Some circuits have their components connected in series, others have a parallel-circuit form. Still others are different combinations of series and parallel circuits. Combination-type circuits are the most widely used arrangements. These arrangements often are a result of switching action or of variable resistance. When such applications need to be constructed or examined, you need a good command of Ohm's and Kirchhoff's laws. One example is shown below.

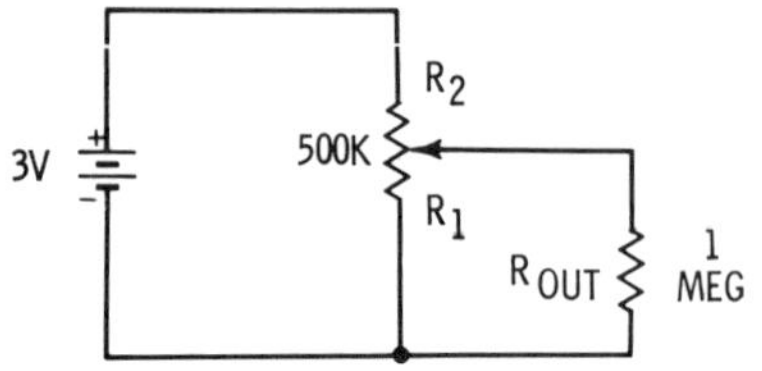


**The potentiometer.**

The output voltage (across $R_{out}$) equals $3V - I_T R_2$, $I_T = 3V/R_T$, and $R_T = \frac{R_1 \times R_{out}}{R_1 + R_{out}} + R_2$. Another method or approach would be to find $IR_{out}$. This would require calculating I.

$$I = I_T - I_{R1}$$

The following diagram is an example of resistor voltage division. The goal in this case is to control the current through the two paths.

**Voltage divider.**

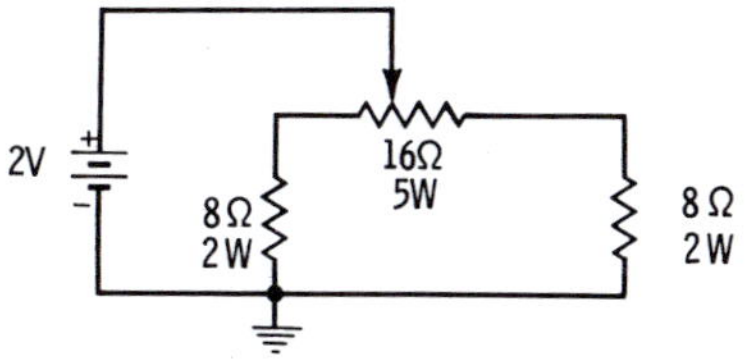


If the wiper of the potentiometer is centered, the resistance from either side to ground will be equal.

The following circuit is another example of a variable voltage output, presenting to the source a constant or near-constant resistance. With the wiper centered on both rheostats, the output voltage is approximately 2.5 volts.

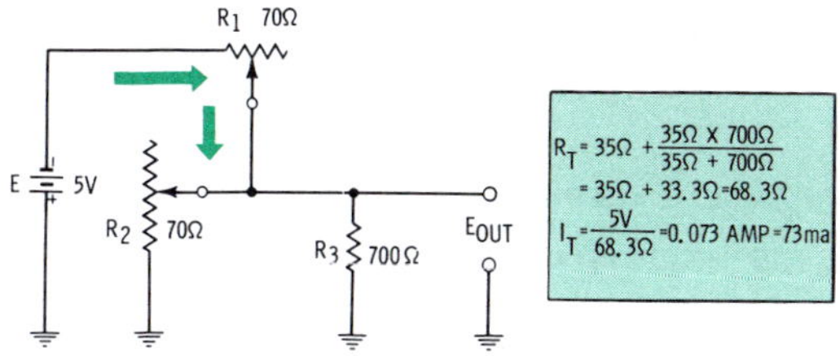


The voltage across $R_1$ will equal 73 ma times 35 ohms, or 2.555 volts. This leaves 2.445 volts for the output. Dividing 2.445 volts by $R_3$ equals approximately 3.49 ma.

**Q22. Solve for the current through the output resistor in the potentiometer circuit on the opposite page, with the wiper arm first at the top of the 500K potentiometer and then in the center (as shown).**

**Q23. What would be the total current in the potentiometer circuit at the top of this page? (Assume the wiper is in the center of the potentiometer.)**

**Q24. What will be the total resistance in the same circuit if the wiper is all the way to the left?**

**Your Answers Should Be:**

**A22.** If the wiper is all the way to the top, the output voltage will be 3 volts. Therefore the current will be equal to the 3 volts divided by 1 megohm of resistance. This results in **3 μa** of current flow. The total resistance across the source at this time will be 333K. If the wiper were in the center position, the total resistance would be 450K. This results in **6.66 μa** of current flow. The current flow through $R_2$ (250K) causes 1.67 volts to be dropped across $R_2$. The resulting output will be 3V — 1.67V, or 1.33V.

**A23.** The total current for the circuit would be 2 volts divided by the total resistance. $I_T = 2V/R_T$ ($R_T$ = 8 ohms plus 8 ohms divided by 2).

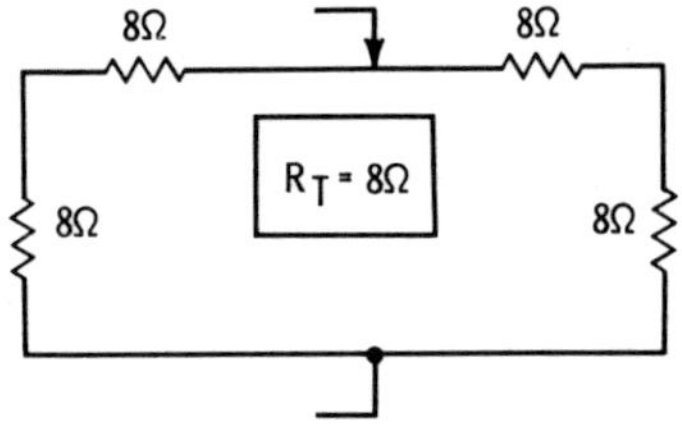


The total current will equal: 2V/8Ω = **0.25 amp**

**A24.** If the arm is all the way to the left, the total resistance will equal: $\frac{8 \times (16 + 8)}{8 + (16 + 8)} = 6\Omega$. $I_T$ for this setting is 2V/6Ω = 0.33 amps.

## WHAT YOU HAVE LEARNED

1. The basic circuit is a series circuit.
2. The series circuit has the same current through all components.
3. The parallel circuit has the same voltage across all branches.
4. IR drop is determined by multiplying the value of a resistance (in ohms) by the value of the current (in amps) flowing through it.

5. In a series circuit the total IR drop across the resistances is equal to the effective total source voltage.
6. The sum of the branch currents in a parallel circuit is equal to the total current of the circuit.
7. Both series and parallel solutions may be employed when solving for current, voltage, or resistance in combination circuits.
8. Kirchhoff's law is an application of Ohm's law.
9. Kirchhoff's law for series circuits states that the sum of all voltages around a circuit, from one point through the circuit and back to the point, will be equal to zero. In addition, the sum of all the voltage drops will equal the effective voltage of the source.
10. Kirchhoff's law for parallel circuits states that the amount of current leaving a junction must be equal to the amount of current entering the junction.
11. The total voltage of the source will be the product of $R_T$ times $I_T$.
12. In a series circuit, the Kirchhoff's law application results in zero if the direction of current is assumed correctly.
13. If the current is assumed to be going in the wrong direction, the result is equal to twice the effective voltage of the source.

# 6

# Electromagnetism

## *What You Will Learn*

**This chapter explains the principles of magnetism for DC applications. It includes a description of magnetism, natural magnets, electromagnets, magnetic properties and relationships, magnetic measurements and associated terms, and DC applications of electromagnetic principles. The DC applications include the effects of magnetic fields on current flow and electrical reactions associated with relays, motors, etc.**

**The fundamental characteristics and typical applications of DC electromagnetism are employed in electrical machinery, radio equipment, laboratory and test equipment, automotive devices, television, radar equipment, computer systems, and many others.**

**When you complete this chapter you will be able to visualize and describe magnetic principles for both permanent magnets and electromagnets, perform experiments and describe the actions and reactions observed, and relate the operating principles of relays, motors, solenoids, and meters to electromagnet fundamentals.**

### HISTORY OF MAGNETISM

During the ancient period in the world's history, in a district in Asia Minor known as Magnesia, the Greeks noticed that a lead-colored stone had an attraction for small particles of iron ore.

In later years the Chinese made use of this stone in their desert travels. They suspended the stone or floated it on water and called it **loadstone,** meaning "leading stone." Loadstone (also spelled **lodestone**) is a natural magnet, because it possesses magnetic properties when found in its natural state. At the present time the most common method of producing electricity is through the use of the magnetic properties of certain materials.

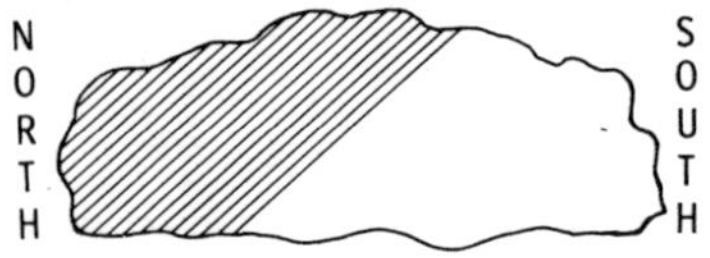


**A loadstone.**

## WHAT IS MAGNETISM?

The dictionary defines magnetism as "a peculiar property possessed by certain materials by which they can naturally repel or attract one another according to determined laws."

In order to provide a better understanding of magnetism it will be necessary to look further into this definition and study the properties, circumstances, and laws referred to.

Magnetism is actually a force which cannot be seen, although you can witness the effects of magnetism on other materials.

## THE MAGNET

We have already discussed one magnet, the loadstone, as being a natural magnet found in the earth. Magnets manufactured today are much stronger than the loadstone.

Iron, cobalt, and nickel are used in the manufacture of artificial magnets. Iron is easy to magnetize, but loses its magnetic properties almost immediately after the magnetizing force is removed. Steel is harder to magnetize, but it holds its magnetism over a greater period of time after the magnetizing force is removed.

The Chinese learned that when the loadstone was suspended or it was floated on a liquid, one end of the stone always pointed in a given direction. Today we know that any magnetic or magnetized material, when suspended or floated, aligns itself with the earth's magnetic field. The end of the magnet or magnetized material that points toward the north pole of the earth is called the "north-seeking" pole or "north pole"; the opposite end is called the "south-seeking" pole or "south pole."

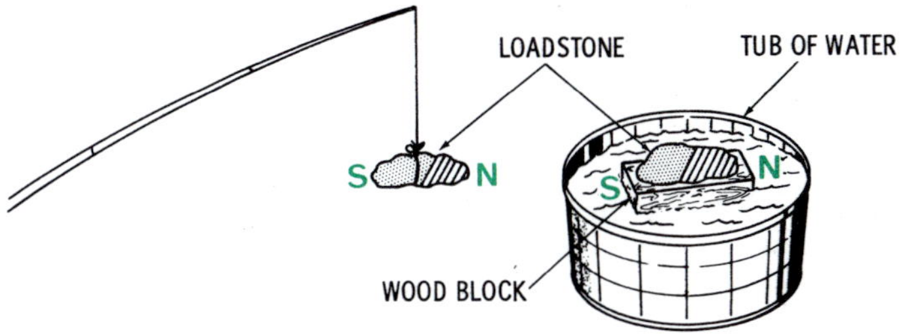


Since magnetism is more pronounced in iron and its alloys than in most other materials, we will take a close look at an atom of iron.

**An atom of iron.**

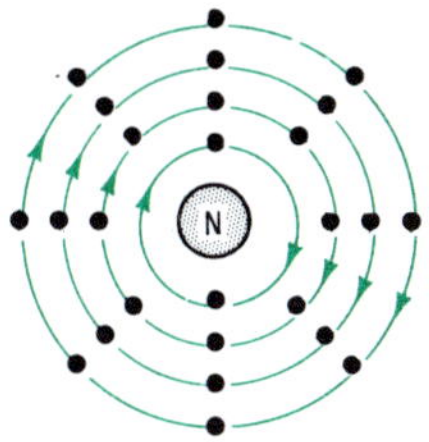


Notice that the majority of electrons in orbit around the nucleus appear to be traveling in the same direction. This is the first clue as to why certain materials are easy to magnetize while other materials are almost impossible to magnetize. If you could take a close look at the atoms in a material that cannot be magnetized, you would see that the electrons in orbit appear to be traveling in different directions; they will cancel out each other's magnetic effects, thus preventing any external magnetic field.

You have learned that any magnet or magnetized material has a north-seeking and a south-seeking pole. One important characteristic possessed by magnets is that if the north poles of two magnets are brought near each other, they repel one another. This also occurs if two south poles are brought near each other.

**Q1. Write the definition of a molecule.**

**Q2. How do like charges affect each other?**

**Your Answers Should Be:**

**A1.** A molecule is the **smallest particle of any substance that still retains the physical characteristics of that substance.**

**A2.** Like charges are **repelled** by each other.

## Magnetic Molecular Alignment

If you were able to view the molecules inside a block of unmagnetized iron, you would see the total disarrangement of the molecules.

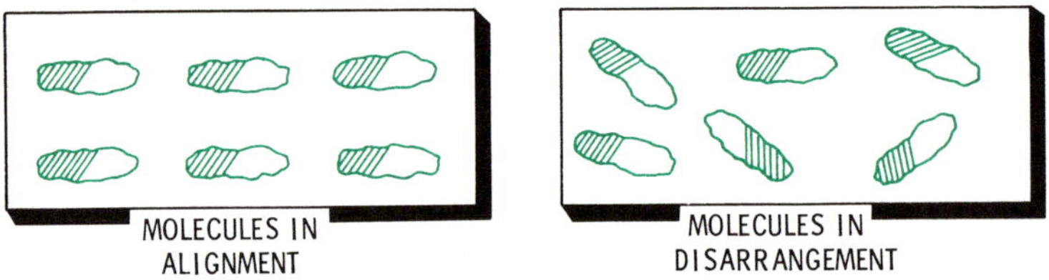


Each molecule within a bar of iron has its own north-seeking and south-seeking poles. Although the magnetic strength of a single molecule is very weak, there are many millions of molecules in a very small piece of metal. When magnetically aligned in the same direction, they can develop a strong magnetic field. This is known as the **molecular theory of magnetism.**

To magnetize a bar of iron, stroke the bar with a material known to be a magnet. Let us assume you choose to apply the north pole of the magnet to the iron. In the illustration below, stroke the iron bar from left to right.

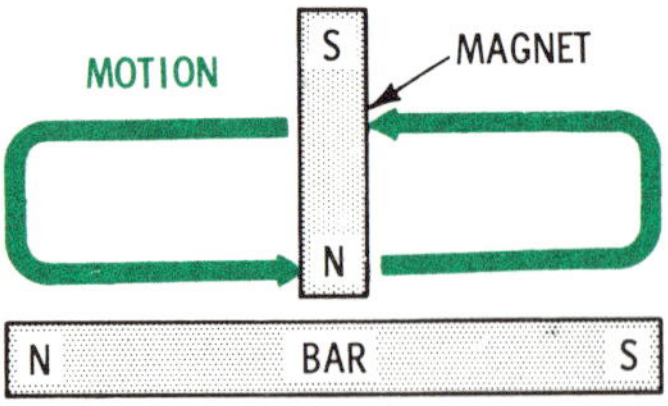


**Magnetizing an iron bar.**

Note that in stroking the iron bar, the same pole of the magnet is always applied to the iron bar and the stroking action is always in the same direction. Make sure the magnet is lifted free of the bar at the end of each stroke.

An alternate method of magnetizing an iron bar is to apply the magnet at the center of the bar and stroke in one direction. After half of the bar is magnetized, reverse the magnet and, again starting at the center, stroke the iron bar in the opposite direction.

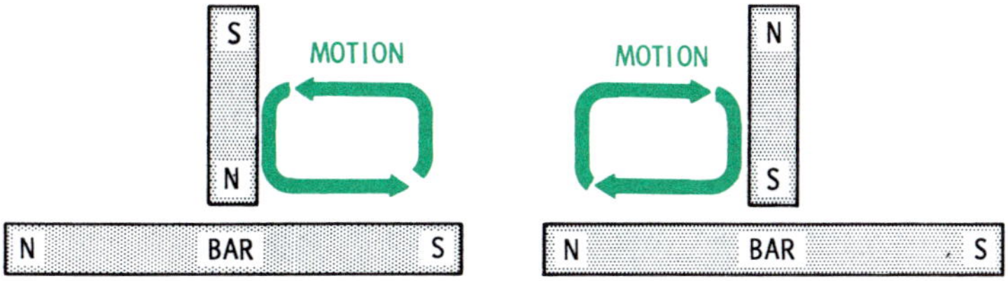


A steel bar can be magnetized in exactly the same way. Steel requires a greater force to align its molecules and therefore takes longer to magnetize. However, steel retains its magnetic properties for a much longer time than iron.

Because steel retains its magnetic properties, it is considered to have a high **retentivity** (the property of any material to remain magnetized).

### Other Methods of Magnetizing Metal

Whenever a piece of iron or steel is placed in a magnetic field, it assumes the properties of the magnetic field.

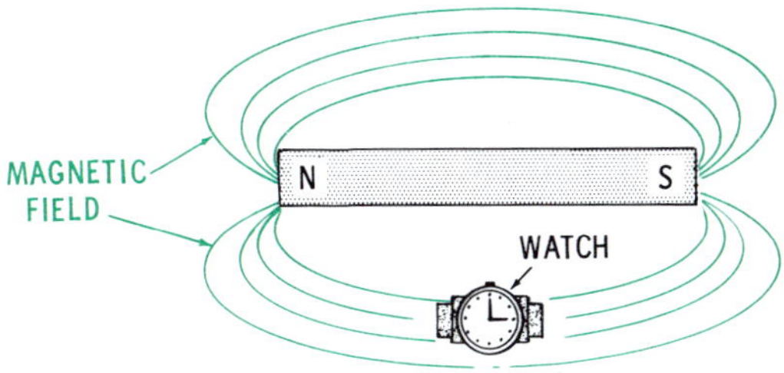


You probably know the danger to your watch if you wear it when working near magnets. Many watches made today are said to be nonmagnetic. This doesn't mean that you can lay your watch on a strong magnet with safety, but it does mean that the watch is shielded from ordinary magnetic fields (lines of force surrounding a magnet).

**Q3. What is the retentivity of iron as compared with steel?**

**Q4. What is meant by a magnetic field?**

**Your Answers Should Be:**

**A3. The retentivity of iron is very low.**

**A4.** A magnetic field is **the pattern of lines of force that surround a magnet.**

### Magnetic Lines of Force

All magnets have invisible force lines surrounding them. These lines leave the north pole of a magnet, form a loop, and enter the south pole of the magnet, completing the loop inside. These loops run parallel to each other inside the magnet and never cross or unite.

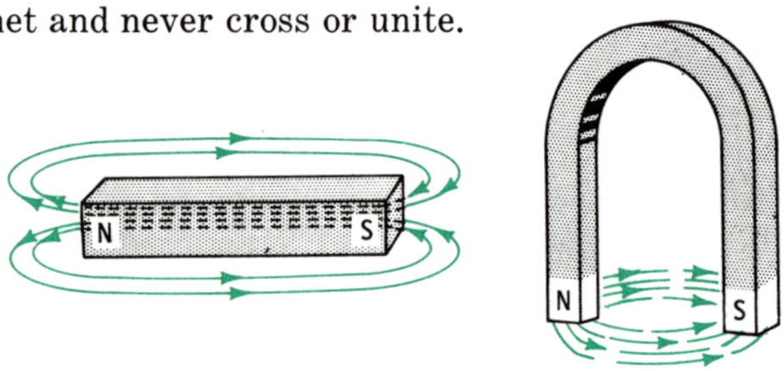


**Arrows show paths of magnetic lines of force.**

The lines formed by the magnetic loops are called **magnetic lines of force.** The area occupied by these lines is called the **magnetic field.** The magnetic field is the induced energy surrounding the magnet or the space through which the influence of these magnetic lines of force can be measured. The strength of the magnetic field is measured by determining the number of magnetic lines of force per unit area surrounding the magnet.

These lines are invisible; therefore, you may wonder how their total number can be determined or what pattern they form. A simple experiment that you may wish to perform will answer these questions and enable you to see these lines for yourself.

### Magnetic Field Pattern Demonstration

You will need:

1. A bar or horseshoe magnet.
2. A piece of glass or clear plastic about 12 inches square.
3. A small can of iron filings.

Place the glass or plastic sheet over the magnet and sprinkle a small amount of iron filings (about a thimble full) over the magnet area on the surface of the sheet. Tap the sheet and notice how the iron filings form a definite pattern similar to that shown in the figure below.

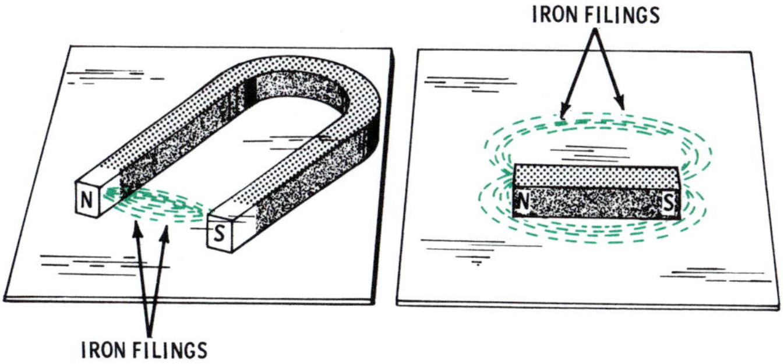


It was stated previously the lines of force were invisible but that you could see their effects. Notice the heavy concentration of iron filings near the poles of the magnet.

It is possible to magnetize an iron or steel bar by stroking it with a magnet. A steel bar or rod can also be magnetized by placing it parallel to the earth's magnetic field and striking it several sharp blows with a hammer. The force from these blows causes the molecules in the bar or rod to change positions and to align themselves with the earth's magnetic field. If a screwdriver becomes magnetized, strike it on a hard surface a few times. Providing its original magnetic properties were rather weak, this striking will rearrange the molecules and demagnetize the screwdriver. Be sure the screwdriver isn't held parallel to the earth's magnetic field as it strikes the hard surface.

Heating, as well as jarring, reduces the magnetism of any material. When iron is heated above 770°C, it can no longer be magnetized or hold any magnetism. Heating a material accelerates the movement of the molecules, and this action causes the molecules to rearrange their alignment.

**Q5. How is the strength of the magnetic field around a magnet determined?**

**Q6. What part of a magnet has the greatest magnetic attraction for a steel bar?**

> **Your Answers Should Be:**
>
> **A5. The number of lines of force per unit area** around a magnet indicates its magnetic field strength.
>
> **A6. The area around either pole** has the greatest influence on a steel bar.

## Magnetic Poles

The path of magnetic lines of force can be controlled. The lines of force concentrated at the poles of the magnet are much closer together than those surrounding the magnet. This is true because magnetic lines of force always take the path of least opposition. Iron or steel offers less opposition to these lines of force than air or other nonmagnetic material. This principle can be used to advantage. If an iron or steel ring is placed around a watch, the magnetic lines will follow a path through the ring and will not pass through the watch. This method of diverting magnetic lines of force is called **magnetic shielding.**

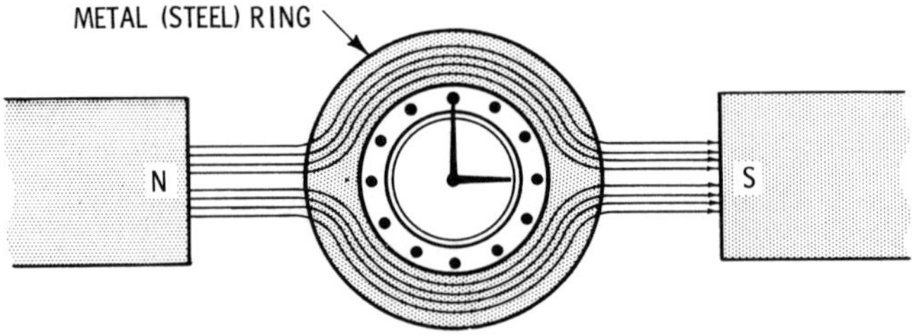


**Magnetic shielding.**

The minimum number of poles a magnet can have is two—a north-seeking pole and a south-seeking pole. It is possible, however, for a magnet to possess more than two poles.

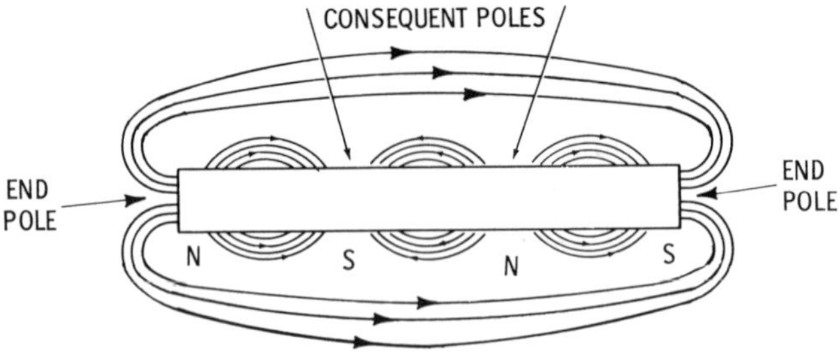

The poles between the ends of a magnet are called **consequent poles.** Notice there are magnetic fields existing between the consequent poles and the end poles. These fields are the same as the field that exists between the end poles. The magnetic lines of force leave a north-seeking pole and enter a south-seeking pole.

## TYPES OF MAGNETS

Basically, there are two types of magnets—permanent and temporary. As their names imply, one magnet retains its magnetism for a long period of time (years in some cases), and the other loses its magnetism almost as soon as the magnetizing force is removed. Manufactured magnets are called artificial magnets since the only natural magnet is the loadstone. Incidentally, a loadstone is very weak compared to a manufactured magnet; therefore, a loadstone has very few applications.

### Applications

The types of magnets which have been discussed are widely used in speakers, meter movements, and magnetic compasses. You may wonder about the third use since a magnet deflects the needle of a compass. A compass installed on most boats, cars, or airplanes is usually surrounded by metal. This metal is affected by the earth's magnetic field. Small bar magnets, called compensating magnets, are placed around the compass to counteract the effects of the earth's magnetic field on the surrounding metal. This makes it possible to use the compass in such places.

**Q7. Magnetic lines of force always take the path of _____ opposition.**

**Q8. Diverting magnetic lines of force is one method of magnetic __________.**

**Q9. What is the minimum number of poles a magnet can have?**

**Q10. The poles between the ends of a magnet are called ___________ poles.**

**Q11. Magnetic lines of force leave the _____ pole and enter the _____ pole.**

**Your Answers Should Be:**

A7. Magnetic lines of force always take the path of **least** opposition.

A8. Diverting magnetic lines of force is one method of magnetic **shielding.**

A9. **Two.**

A10. The poles between the ends of a magnet are called **consequent** poles.

A11. Magnetic lines of force leave the **north** pole and enter the **south** pole.

## Horseshoe Magnets

The magnets used in meters are shaped like a horseshoe. By bringing the two poles close together, the lines of force are concentrated and thus provide a much stronger magnetic field.

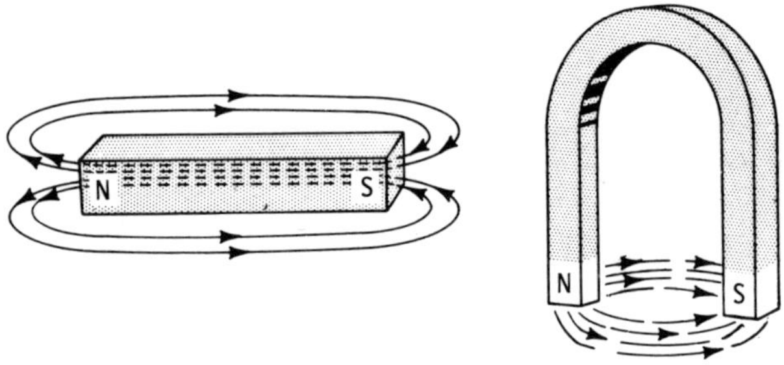


## Care of Magnets

Magnets that are not properly cared for lose their magnetic properties over a period of time. How much magnetism is lost depends on many variables—how the magnet was originally magnetized, how it is used, where it is used, etc. When a horseshoe magnet is not in use, a soft iron bar should be placed across the poles. This bar will provide a path for the magnetic lines of force, and the magnet will retain its magnetic properties for a much longer period. The iron bar used for this purpose is called a **keeper.** Bar magnets should be stored parallel to each other with unlike poles together.

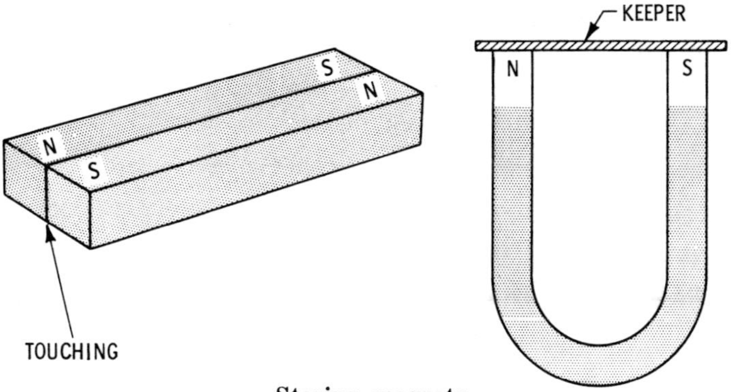

**Storing magnets.**

In some cases, it is important for the magnet to maintain a specific magnetic force over a long period of time. When this function is necessary, magnets are put through a process of **aging** during their manufacture. This is done by placing the magnet in an oven and subjecting it to controlled temperature changes and to vibrations. This process causes the strength of the magnet to remain nearly constant for a long period of time.

**Reluctance**

Some materials offer more opposition to magnetic lines of force than do others. In magnetic circuits this opposition is called **reluctance.**

In the study of electrical circuits, you learned that electromotive force causes current to flow in a circuit and the flow of that current is limited by resistance. There is also a magnetic circuit in which the magnetic lines of force form closed loops, called **flux loops.** The force that produces these flux loops is called the **magnetomotive force (mmf).** The opposition to the flux loops is called **reluctance.** Notice the similarity to the electrical circuit. In a magnetic circuit the magnetic lines of force always take the path of least reluctance. This is why the magnetic lines followed the steel ring around the watch discussed previously. The steel ring offered less reluctance than the air and the nonmagnetic metal of the watch in the center of the ring.

**Q12. The opposition that some materials offer to magnetic lines of force is called __________.**

**Your Answer Should Be:**

**A12.** The opposition that some materials offer to magnetic lines of force is called **reluctance.**

## Magnetic Flux

An expression for determining the amount of flux present in a magnetic circuit is:

$$\text{flux} = \frac{\text{magnetomotive force}}{\text{reluctance}}$$

Flux varies directly with the magnetomotive force and inversely with the reluctance. This is the Ohm's law expression for magnetic circuits. Compare the two formulas.

$$\text{Current} = \frac{\text{electromotive force}}{\text{resistance}}$$

Magnetic flux is the total number of magnetic lines existing in a magnetic circuit or extending through a specific region. The symbol for magnetic flux is the Greek letter $\phi$ (phi). One magnetic line of force is equal to 1 **maxwell.**

The concentration of these magnetic lines determines the **flux density.** The symbol for flux density is **B,** and the unit of measurement is the **gauss.** One gauss is a flux density of one line of force per square centimeter.

The degree of flux density between the poles of a horseshoe magnet is directly proportional to the area of the air gap between the poles. The force of attraction or repulsion between the poles varies directly with the strength of the poles and inversely with the square of the distance separating them. This force can be determined as follows.

$$F = \frac{P_1 \times P_2}{\mu d^2}$$

where,

F is the force between poles in dynes (unit of force),
$P_1$ and $P_2$ are the strengths of the two poles,
d is the distance in centimeters between the poles,
$\mu$ is a constant that depends on the medium between the poles. It is 1 for air and greater than 1 for other mediums.

To find the total number of flux lines, multiply the flux density (in gausses) by the area (in square centimeters).

Certain materials have more opposition (reluctance) to magnetic lines of force than others. It is therefore true that some materials allow magnetic lines of force to pass more easily than others. The ease with which magnetic lines of force pass through a material is known as **permeance**, the reciprocal of reluctance (permeance $= \frac{1}{\text{reluctance}}$). Any substance that allows the magnetic flux to pass with little or no opposition is said to have a high **permeability.** Iron, for example, has a high permeability. High-permeability materials can be easily magnetized, but they will not retain their magnetism. Permeability varies with the intensity of the magnetic field in which the material is located.

It is also possible to determine the flux in any material by multiplying the magnetomotive force by the permeance.

$$\text{flux} = \text{mmf} \times \text{permeance}$$

Not all materials can be magnetized. Actually, materials can be broken down into three classifications—diamagnetic, paramagnetic, and ferromagnetic. **Diamagnetic** materials are those that normally cannot be magnetized. **Paramagnetic** materials are those that are difficult to magnetize. **Ferromagnetic** materials are those that are relatively easy to magnetize. Some ferromagnetic materials are iron, cobalt, nickel, silicon steel, and cast steel.

A diamagnetic material is extremely difficult to magnetize and has a permeability of less than 1. A paramagnetic material is also difficult to magnetize, but has a permeability of slightly greater than 1. A ferromagnetic material is easily magnetized, and its permeability is quite high.

**Q13. If there are 3,000 magnetic lines of force passing through a magnet, what is the magnetic flux?**

**Q14. If there are 2,700 maxwells in a cross-sectional area of 9 square centimeters and the lines are evenly spaced, how many maxwells are there in 1 square centimeter?**

**Q15. In an area of 5 square centimeters the flux density is 6,000 gausses. How many flux lines pass through the area?**

**Your Answers Should Be:**

**A13.** The magnetic flux is **3,000 maxwells.**

**A14.** There are **300 maxwells** in 1 sq. cm.

**A15.** **30,000** flux lines pass through the area.

## ELECTROMAGNETS

There is another type of magnet that has a wide range of applications in electricity. This is the **electromagnet.** The dictionary defines an electromagnet as "a bar of soft iron that will become a temporary magnet if an electrical current is caused to pass through a wire coiled around it."

Electromagnetism was first discovered by Hans C. Oersted, a Danish scientist, in 1820. Oersted found that a needle placed near a wire would deflect when current passed through the wire. Further experiments led to the discovery that current flowing through a conductor creates a magnetic field about the conductor. This magnetic field is composed of lines of flux (magnetic lines) that encircle the conductor at right angles to the flow of current. The flux lines are uniformly spaced along the length of the conductor.

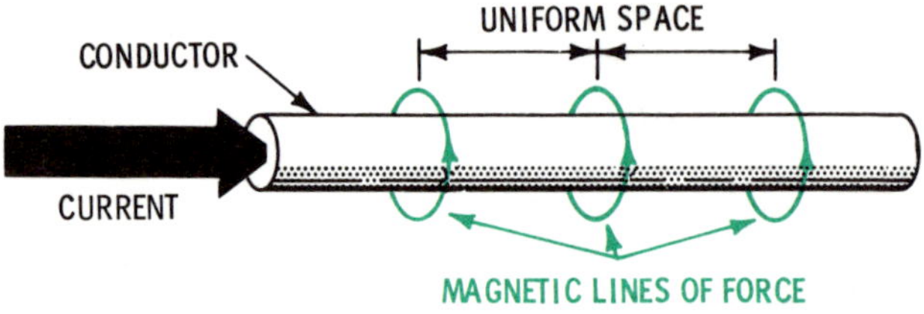


This is another method of creating a magnet. The magnetic field around a straight conductor is not very strong, but it exhibits the same properties as the bar or horseshoe magnet discussed previously. If a compass is placed near the conductor, the compass needle is deflected at right angles to the current-carrying conductor. The compass also indicates that the magnetic field around the current-carrying conductor is polarized. If the current is reversed through the coil, the position of the compass needle also reverses.

The magnetic field around a conductor diminishes with an increase in distance from the conductor.

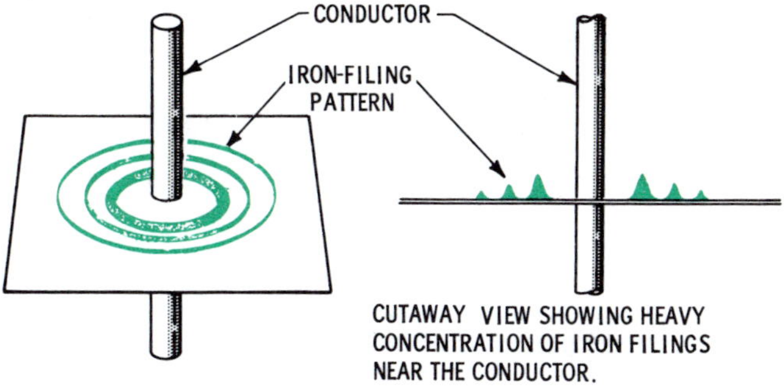


**Field is strongest near the conductor.**

## Left-Hand Rule

To determine the direction of the magnetic field around a current-carrying conductor, grasp the conductor in your left hand, with your thumb pointing in the direction of current flow. The direction that your fingers curl around the wire indicates the direction of the magnetic field.

**Left-hand rule.**

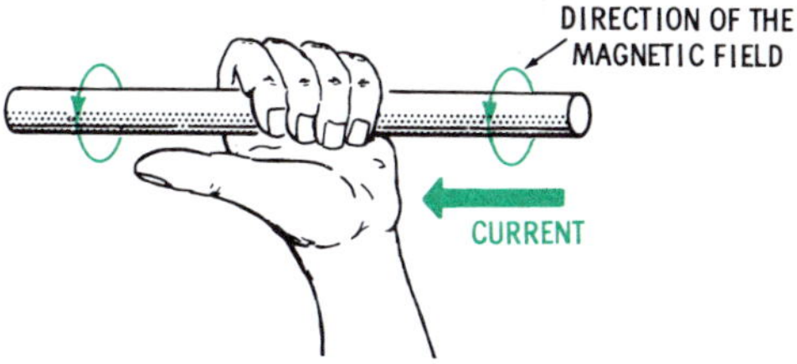


Do not attempt such an experiment on a bare wire. Instead, place your left hand in a position away from the wire with your thumb pointing in the direction of current flow. Your curled fingers will indicate the direction of the magnetic field. It isn't necessary to grasp the conductor.

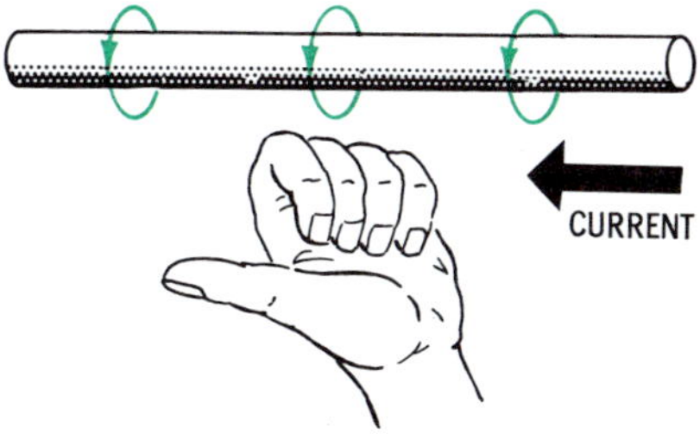


**A16. In what plane do the magnetic lines around a current-carrying conductor lie?**

**Your Answer Should Be:**

**A16.** The magnetic lines around a current-carrying conductor lie in a plane at **right angles to the conductor.**

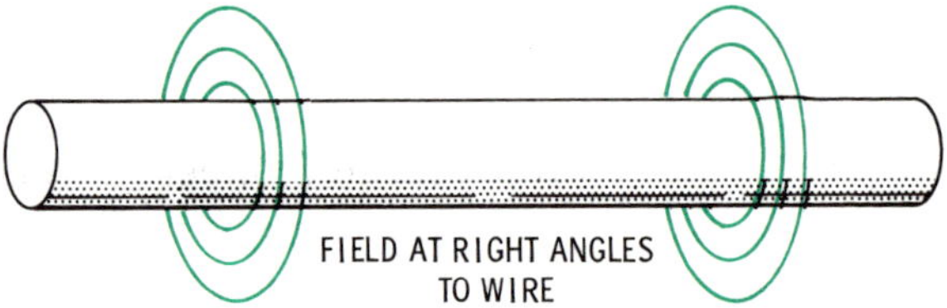


## Magnetic Field Strength

Magnetomotive force is the force that tends to drive the flux through a magnetic circuit. The unit of magnetomotive force is the **gilbert.**

The unit of measurement used to express field intensity is the **oersted.** The strength of the magnetic field can be found by using the following expression:

$$H = \frac{I}{5d}$$

where,

H is the field intensity at a point nearest the wire in oersteds,
I is the current through the wire in amperes,
d is the distance of this point from the axis of the wire in centimeters (1 inch = 2.54 centimeters).

## Constructing an Electromagnet

The magnetic field developed around a straight wire or conductor is seldom strong enough to be useful. However, if the wire is formed into a coil, the magnetic field becomes quite strong.

The figure on the next page shows the action that takes place when current flows through a coil. All of the magnetic lines of force enter the coil at one end and emerge at the opposite end.

The strength of the magnetic field is directly proportional to the number of turns in the coil and the current passing

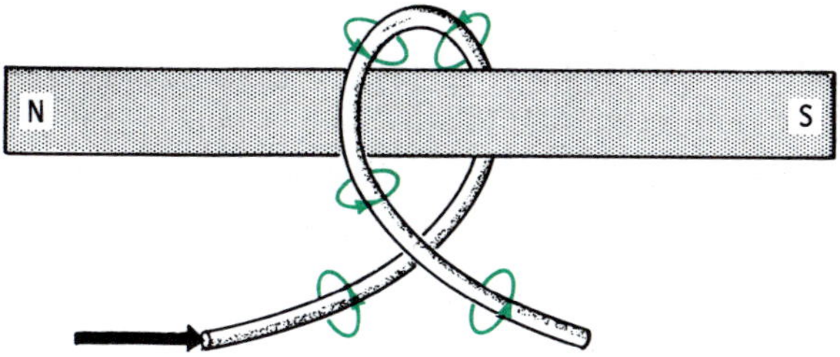


through them. A coil with a large number of turns has a magnetic field of greater strength than one with a small number of turns. The magnetic field is greater when a larger current flows through the coil.

Since all magnetic lines of force form a loop, poles similar to those of a permanent magnet are established on the coil. The poles form at each end of the coil and their polarities depend on the direction of current flow.

The left-hand rule is employed to determine the magnetic polarity. Grasp the coil in your left hand with your fingers pointing in the direction of current flow. Your thumb points in the direction of the north-seeking pole of the coil.

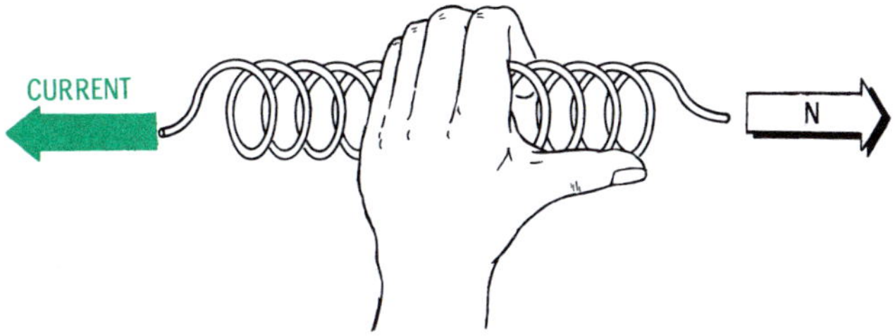


It was stated previously that the flux density is much greater in a block of iron than it is in air. Therefore, if an iron core is added to the current-carrying coil, the magnetic loops will concentrate through the core, increasing the flux density and strength of the electromagnet. The magnetic lines around the coil are called **induction lines.** Soft iron cores are used in electromagnets because of the high permeability of iron.

**Q17. What is the field intensity at a distance of 5 inches from the center of a wire carrying 100 amps?**

**Q18. Why is the magnetic field around a coil stronger than the magnetic field around a straight wire?**

**Your Answers Should Be:**

**A17.** The field intensity is **1.57 oersteds.**

$$H = \frac{I}{5d} = \frac{100}{5 \times 12.7 \text{ cm}} = \frac{100}{63.5} = 1.57 \text{ oersteds}$$

**A18.** When a straight wire is formed into a coil, the magnetic lines around each turn are reinforced.

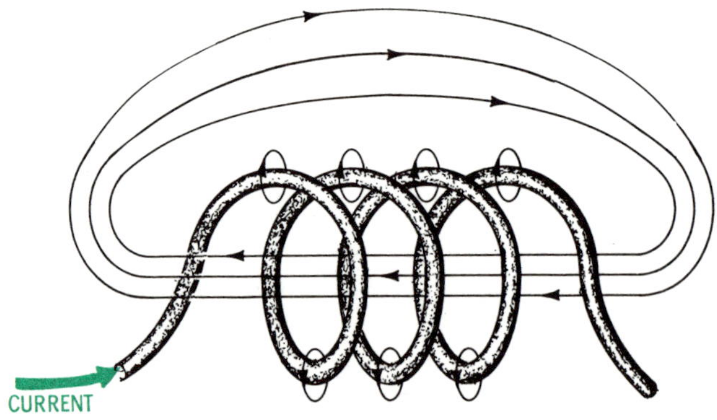


The strength of an electromagnet can be determined by connecting a coil across a battery and placing an iron rod, suspended by a small hand scale, near the coil. When the circuit is energized, current flows through the coil and the magnetic field that is developed attracts the iron rod. The amount of pull can be read directly on the hand scale after subtracting the weight of the rod.

If the overall length of a coil is less than its diameter, the strength of the field can be calculated by the following expression.

$$H = \frac{2\pi NI}{10r}$$

where,

H is the field intensity in oersteds,
N is the number of turns in the coil,
I is the current through the coil in amperes,
r is the radius of the coil in centimeters,
$\pi$ is a constant equal to 3.14.

The two main factors that determine the strength of an electromagnet are the current and the number of turns in the coil. The magnetic field can be varied by altering either factor. The combination of these two factors (I and N) is called **ampere turns.** An electromagnet with 200 turns of wire through which 1 amp is flowing has a field strength equal to an electromagnet with a 10-turn coil through which 20 amps is flowing. In both cases the number of ampere turns is 200.

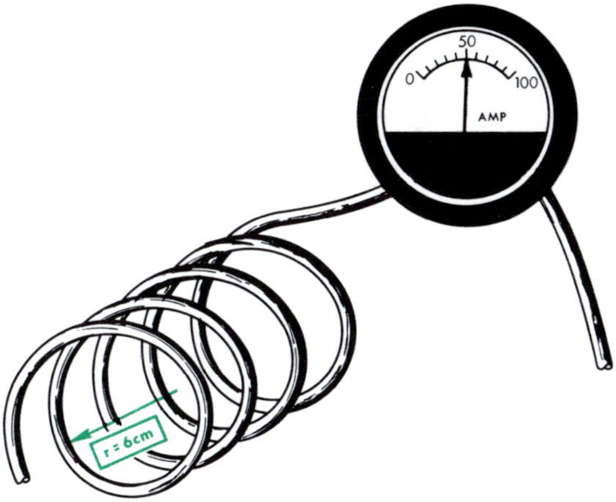


**Magnetic strength depends on current and number of coil turns.**

**Q19. What can be added to a coil of wire to make it a stronger electromagnet?**

**Q20. What is meant by permeable material?**

**Q21. Calculate the field strength of the coil shown in the figure on this page.**

**Q22. What determines the field strength of an electromagnet?**

**Q23. What is the field strength of a coil having 10 amps of current flowing through it, if the coil has a radius of 2 inches and contains 26 turns?**

**Q24. Find the current flowing in a coil having a field strength of 25 oersteds, 15 turns, and a radius of 2 cm.**

**our Answers Should Be:**

A19. Adding **an iron core** to a coil increases the strength of an electromagnet.

A20. Permeable material is **any material that can be easily magnetized.**

A21.

$$H = \frac{2\pi NI}{10r} = \frac{6.28 \times 4 \times 50}{10 \times 6} = \frac{1{,}256}{60} = \mathbf{20.93\ oersteds}$$

A22. The **current** and the **number of turns** in the coil.

A23.

$$H = \frac{2\pi NI}{10r} = \frac{6.28 \times 26 \times 10}{10 \times 2 \times 2.54} = \mathbf{32.14\ oersteds}$$

A24.

$$H = \frac{2\pi NI}{10r}$$

$$2\pi NI = H \times 10r$$

$$I = \frac{H \times 10r}{2\pi N} = \frac{25 \times 10 \times 2}{6.28 \times 15}$$

$$\mathbf{I = 5.30\ amps}$$

**Magnetomotive Force**

Magnetomotive force is defined as the force that produces a magnetic field, and is measured in **gilberts.** In the design of an electromagnet for a particular application, it is often desirable to determine just how much magnetomotive force is required to create a magnet with a specific field strength.

A convenient method of determining the magnetomotive force in a current-carrying air-core coil is to use the following expression.

$$\text{mmf} = 1.257 \times I \times N$$

where,

mmf is the magnetomotive force in gilberts,
I is the coil current in amps,
N is the number of coil turns.

The reluctance of air is 1.257 (this number will be different for an iron-core coil). As can be seen, magnetomotive force is directly proportional to the ampere turns.

### Residual Magnetism

When an electromagnet is de-energized, the magnetic field collapses, but a slight amount of magnetism remains in the core material. This is called **residual magnetism.**

When the magnetic lines of force surrounding the coil concentrate inside the center of the coil, the force magnetically aligns the molecules in the core material. This is similar to the process of magnetizing a metal bar. If the core material is a bar of steel, the results will be different from those for a bar of iron. Once the molecules are aligned in the steel bar, they tend to remain aligned. The core will then retain considerable residual magnetism after the current has ceased to flow through the coil.

### Hysteresis

If the current is reversed in an electromagnet (perhaps many times a second), the magnetic field and the direction of polarization will also reverse. If the core material possesses any residual magnetism, the polarity change in the magnetic field will be somewhat delayed beyond the time when the current is reversed. It is necessary to overcome the residual magnetism before the core can be magnetized in the reverse direction.

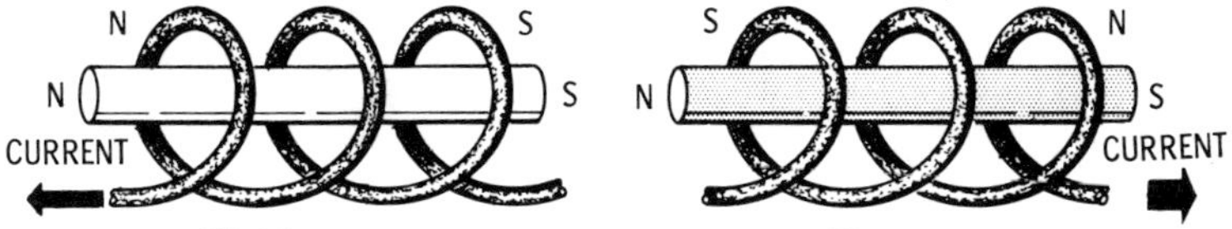


*Fields aiding.* *Fields opposing.*

**Current in an electromagnet.**

When current flows through the coil in the direction shown, the north pole of the magnet is on the left, and the south pole is on the right (left-hand rule for a coil). The magnetic polarity of the core material is identical to the polarity of the coil. When the circuit is de-energized, the magnetic field collapses. Any residual magnetism remaining in the circuit retains its original polarity.

**Q25. What is the magnetomotive force if 2 amperes flows through an 8-turn air-core coil?**

**Q26. What makes the best core material for an electromagnet?**

**Your Answers Should Be:**

**A25.** The magnetomotive force is **20.112 gilberts.**

$$\text{mmf} = 1.257 \times I \times N = 1.257 \times 2 \times 8 = 20.112 \text{ gilberts}$$

**A26.** **Soft iron** makes the best core material because its molecules tend to rearrange themselves easily after the magnetizing force is removed.

If the current through a coil is reversed, the magnetic field also reverses. Before the reverse magnetic field can build up, however, it must first overcome the residual magnetism in the core material of the coil. The residual magnetism opposes the new field, so it is first necessary to reduce the residual magnetism to zero before the new field can be developed. Instead of the magnetic field being developed immediately as the current increases, there is a slight delay. The magnetic field lags the current slightly. This lag is called **hysteresis.**

Energy is required to align the molecules in the core material. If the current through the coil is reversed frequently, considerable energy is required to realign the molecules first in one direction and then in the other. This energy is lost in the form of heat and is called **hysteresis loss.**

The hysteresis lag becomes quite evident when a curve of magnetizing force (H) is plotted against flux density (B).

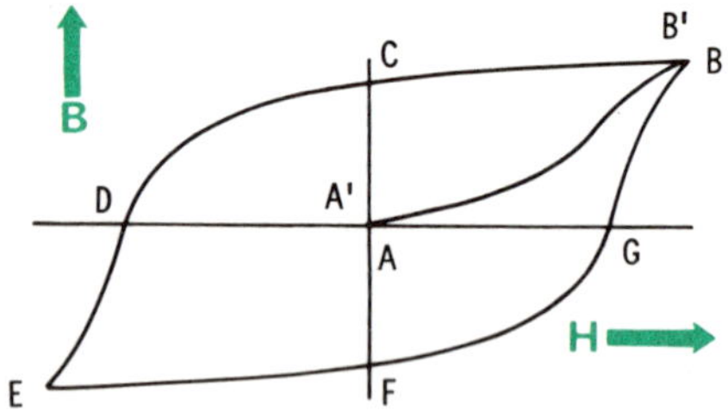


**B-H curve.**

The figure above is referred to as the **B-H curve.** When current flows through a coil, the magnetic field around the coil builds up, as indicated by line A-B. When the current reaches its maximum level, the magnetic field also reaches maximum intensity, as indicated by point B. When the current ceases to flow, the magnetic field collapses along line B-C. When the current reaches zero, the amount of residual magnetism remaining in the circuit is indicated by line A-C.

If the current is reversed in the circuit, the residual magnetism must first be overcome. It falls to zero, as indicated by line C-D. The magnetic field then builds up in the opposite direction along line D-E, reaching maximum concentration at point E. When the current stops flowing in the reverse direction and falls to zero, the magnetic field collapses along line E-F. The distance between points A and F indicates the amount of residual magnetism remaining in the core at that time. If the current were to flow in its original direction, the magnetic field would collapse to zero, as shown by line F-G, and build up to its maximum level along line G-B.

**Saturation**

The magnetic field develops gradually as the current increases. There is a point, however, where the core material cannot accommodate additional lines of flux. When this point is reached, the core is said to be **saturated.** Any additional lines of force will have to flow through the air surrounding the core material. Since the reluctance of air is quite high compared to the reluctance of iron, it is difficult to increase the flux density beyond core saturation.

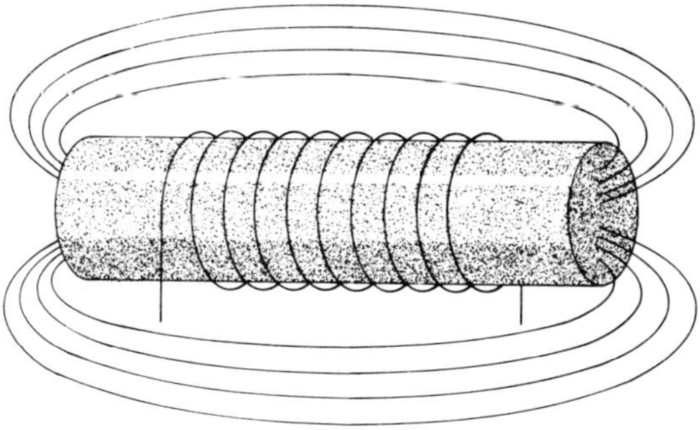

CORE MATERIAL SATURATED

**Magnetic saturation.**

**Q27. Will residual magnetism have any effect on the temperature of the core?**

**Q28. What would the effects be if the core material were removed?**

**Your Answers Should Be:**

A27. **Yes,** the temperature will definitely be affected. When the molecular action increases, the material becomes hot, and energy is expended in the form of heat.

A28. If there were no core material **the B-H curve would increase and decrease in a linear fashion** and there would be no residual magnetism.

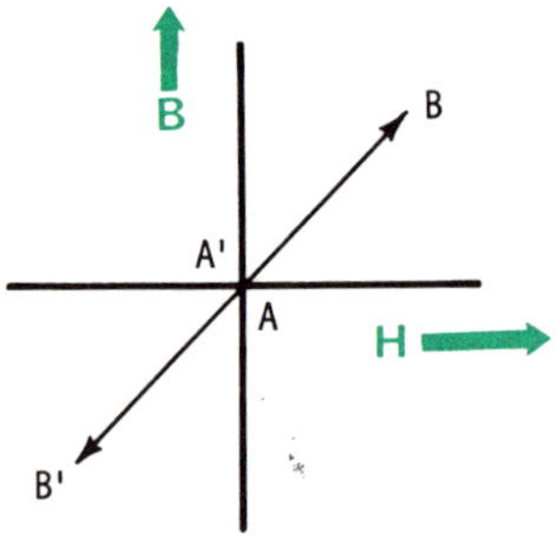


### Magnetic Permeability

Compare the B-H curve on the preceding page with the curve above. Notice that the first curve does not vary in a linear fashion along line A-B or A′-B′. This is true because there is a slight variation in the process of core magnetization. Theoretically, the magnetic field gradually builds up by a definite quantity whenever the magnetizing force is varied by a specific amount.

For example, if the magnetizing force increases by 3 oersteds, the magnetic flux increases by 10,000 maxwells. This is true anywhere along the theoretical curve. In actual practice, however, there is a slight variation. At certain points along the curve an increase of 3 oersteds in the magnetizing force may increase the magnetic flux by only 9,800 maxwells. At other points on the curve, such an increase in the magnetizing force may increase the magnetic flux by 10,300 maxwells. This variation accounts for the nonlinearity of line A-B on the B-H curve.

The following graph compares magnetizing force and flux density when a steel bar is magnetized.

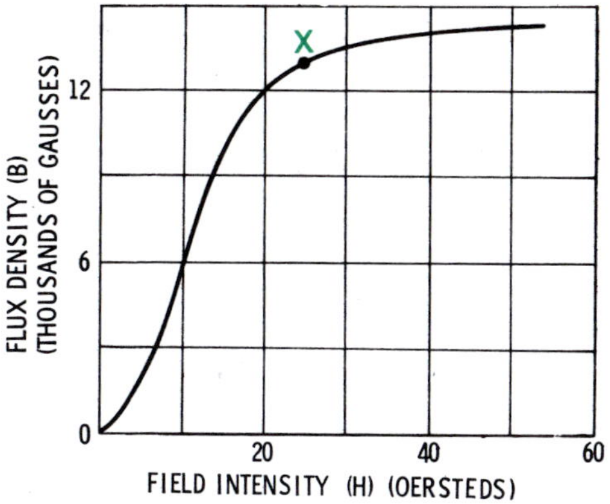


The flux density increases rapidly with only a slight increase in the magnetizing force when the core material is at a point of low field intensity (only a few thousand magnetic lines of force). At a point of high field intensity, a large change in the magnetizing force is required to cause a small change in flux density. Point X indicates the point of saturation.

Permeability can be determined by the following expression.

$$\mu = \frac{B}{H}$$

where,

$\mu$ is the permeability (has no unit of measure),
B is the flux density in gausses,
H is the magnetizing force in oersteds.

When the permeability of a material is low, the reluctance is high, requiring a large magnetizing force to increase the flux density. This can be seen from the following expression for determining flux.

$$\phi = \frac{mmf}{R}$$

where,

$\phi$ is the flux lines in maxwells,
mmf is the magnetomotive force in gilberts,
R is the reluctance.

**Q29. Referring to the graph above, would the permeability at the point of saturation be high or low?**

Your Answer Should Be:

**A29.** The permeability of the core material at saturation would be very **low.**

## Solenoids

A coil wound in the shape of a cylinder or tube is called a **solenoid.** A solenoid is often provided with a movable iron core, or plunger. In this arrangement, the iron core is pulled into the coil when current flows through the turns. Thus, the core can be used to mechanically move some device.

Solenoids are commonly used in relays or circuit breakers. The magnetic field built up in the center of the coil pulls the core into the solenoid, thereby breaking or making the relay contact(s).

**A solenoid.**

## Toroids

Another type of coil that is used in some applications is the **toroid,** which has a ring-shaped core on which the turns of wire are wound to form a complete circle. This design concentrates all the lines of force inside the ring. With all the flux inside the ring, the toroid has no external polarity.

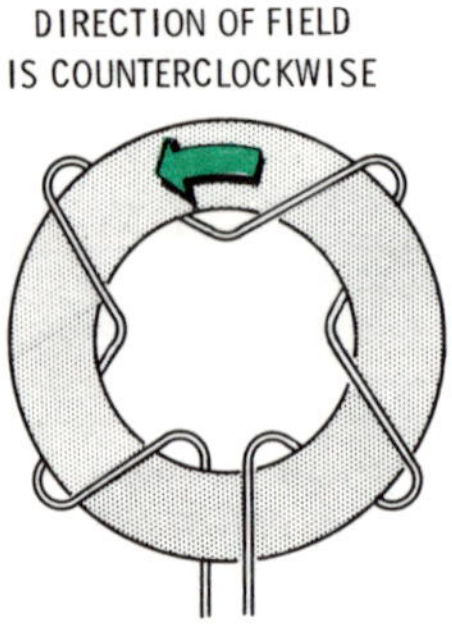


**A toroid.**

## Polarized Electromagnets

A polarized electromagnet has a permanent magnet as its core, as shown in the figure below.

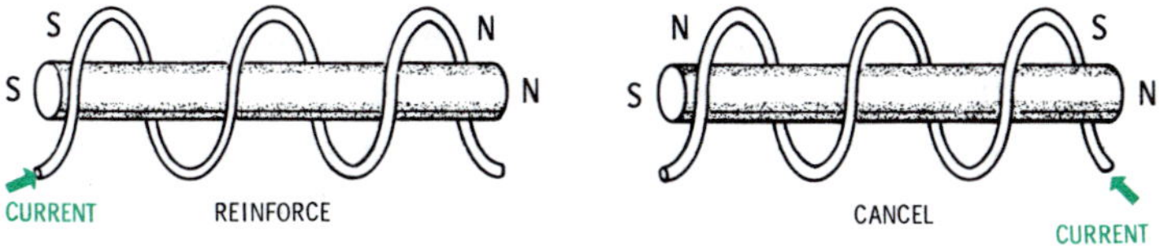


When current flows in the coil, the electromagnet will either add to, neutralize (cancel out), or subtract from the magnetic field of the permanent magnet. Polarized electromagnets are used in telephone and telegraph circuits. The figure below shows a diagram of a telephone-bell circuit.

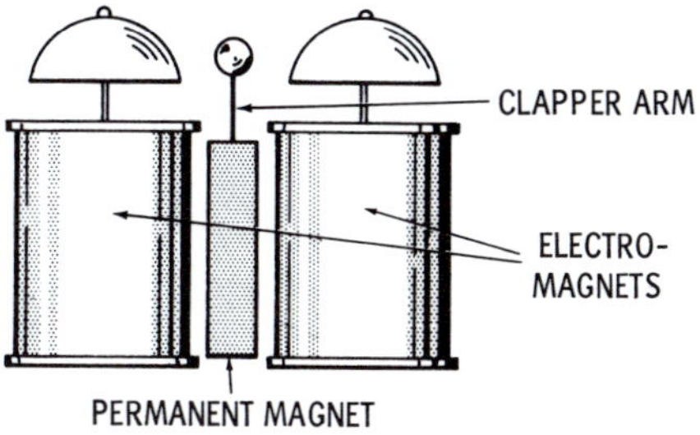


**A polarized electromagnet.**

The clapper arm, which extends down through the center of the electromagnet, is attached to a permanent magnet. The permanent magnet holds the clapper arm in a neutral position between the bells and provides the clapper with a specific polarity. Assume this position holds the clapper arm to the south-seeking pole of the permanent magnet. When current flows through the series-connected coils, the electromagnetic field thus developed adds to the field of the permanent magnet. This combined field pulls the clapper arm to the right, causing the clapper to strike the right-hand bell. When the current is reversed, the magnetizing force of the electromagnet subtracts from the force of the permanent magnet, and the clapper strikes the left-hand bell.

**Q30. How would you determine which is the north-seeking pole of a solenoid?**

**Q31. In what position will the clapper arm be when current is not flowing through the electromagnet?**

**Your Answers Should Be:**

**A30.** The magnetic field around a solenoid is similar to that around any coil. The **left-hand rule** used to find the polarity of a coil may also be used to determine the polarity of a solenoid.

**A31.** The permanent magnet returns the clapper arm to **the center position** when the current stops flowing.

## USES FOR MAGNETS

You may ask why the permanent magnet does not become demagnetized when the electromagnetic field opposes it. Once a permanent magnet becomes magnetized, a strong force is required to disarrange its molecules. The electromagnetic field might be strong enough to do this if it remained for a very long period of time. However, the current through the coil in a specific direction lasts for only a brief period and does not noticeably change the strength of the permanent magnet.

When current passes through parallel wires, the magnetic fields around these wires interact. If the currents flow in the same direction, the fields oppose each other between the wires and aid beyond the wires. This tends to move the wires together. If the currents flow in opposite directions, the wires tend to move apart.

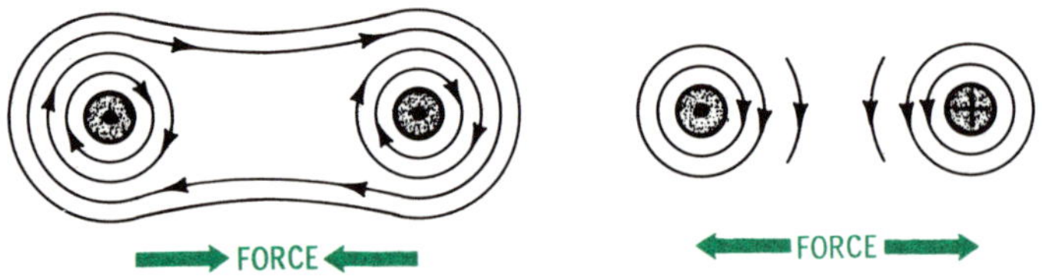


**Parallel wires carrying current.**

### Electric Motors

The principle of attraction and repulsion just shown is used in electric motors and generators. Electric motors are used to provide a mechanical power output from an electrical input. Generators provide an electrical output from a mechanical input.

The force exerted on an electron in a magnetic field is at right angles to the magnetic field. When the electron is placed in both an electric and a magnetic field, the force exerted on the electron is perpendicular to both fields. A right-hand rule is used to determine the direction of force on electron flow in a magnetic and electric field.

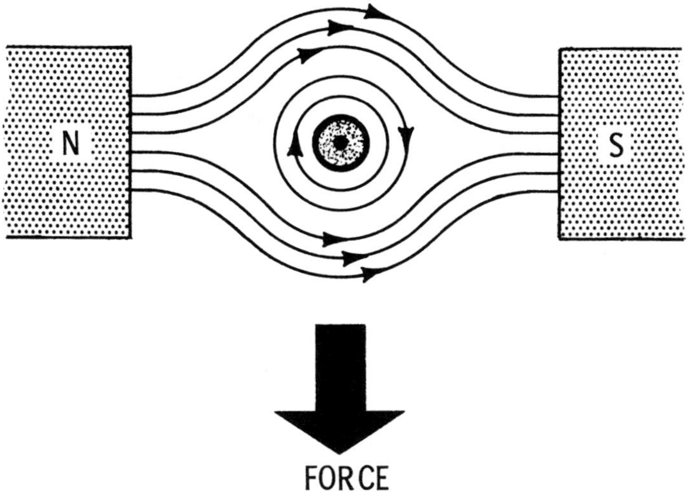


**Force exerted on a current-carrying conductor in a magnetic field.**

The magnetic field around the conductor in the above figure is clockwise. The current appears to be coming out of the page. The direction of the magnetic field of the permanent magnet is from the north-seeking pole to the south-seeking pole, or from left to right in the figure above. Notice that the lines above the conductor and the lines around the conductor are going in the same direction, reinforcing the field above the electron path. Below the conductor the fields are opposing each other.

**Q32. In the figure above, the field around the conductor and the magnetic field of the permanent magnet are opposing each other at a point below the electron path. What effect does this have on electron flow?**

**Your Answer Should Be:**

A32. This weakens the fields and electron flow is forced in a downward direction.

## Right-Hand Rule

Arrange the thumb, index finger, and middle finger of your right hand as shown in the figure below. Point the index finger in the direction of magnetic flux and the middle finger in the direction of electron flow. The thumb indicates the direction of magnetic force on the electron (direction that the wire is repelled).

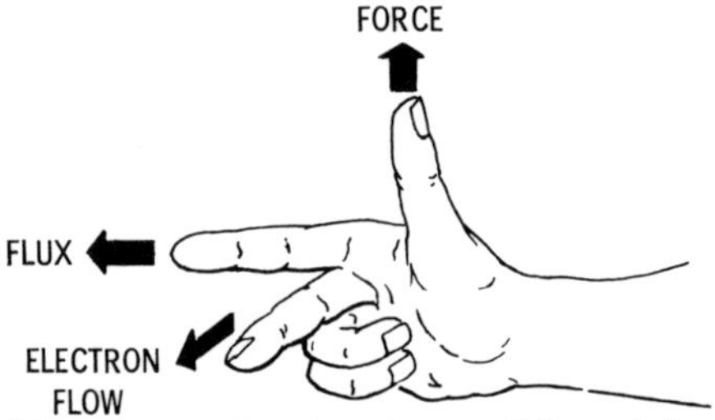


**Right-hand rule.**

If a loop of wire is positioned in a permanent magnetic field, a force acts on the wire each time current passes through it. This force causes the loop of wire to rotate, if it is free to turn, as shown in the figure below.

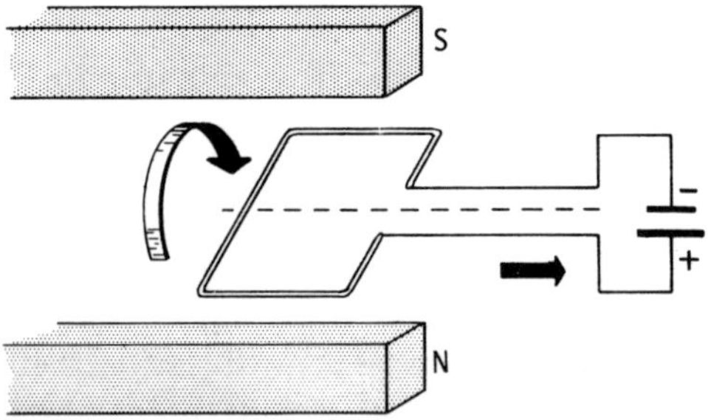


The simple motor shown in the figure above is not very practical. The coil cannot rotate very far because the current always moves through the wire in the same direction. When opposing poles appear opposite each other, the loop stops. Furthermore, any permanent connections to the loop of wire will not allow it to rotate very far.

To overcome these objections, the loop is terminated in two contacts that rotate with the loop. These contacts form the **commutator** of the motor. Electrical connections are made by carbon brushes pressing against the commutator.

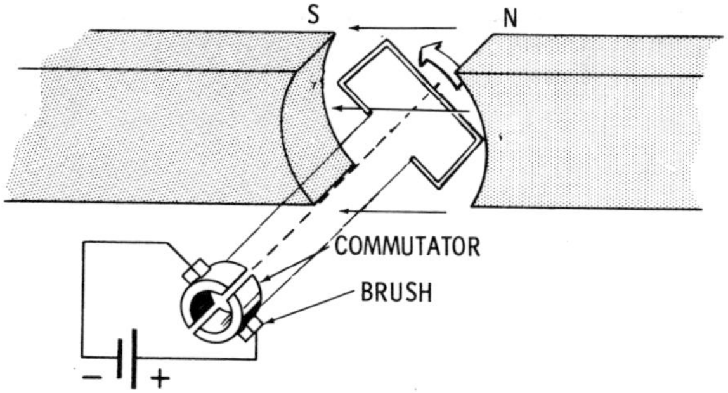


**Motor commutation.**

When current flows through the wire loop, a magnetic field is set up so that the north-seeking pole appears above the loop and the south-seeking pole below. (Check this by employing the left-hand rule for a coil.) The magnetic poles thus created around the loop are attracted by the opposite poles of the permanent magnet; this causes the loop to rotate in a counterclockwise direction. (According to the right-hand rule, the force is downward on the left side of the loop and upward on the right side.)

The loop and commutator rotate together. When the loop has reached a position such that the opposing poles of the electromagnet are adjacent, the commutator will have rotated to a position where the applied voltage is reversed. The current through the loop will now reverse directions, reversing the magnetic field around the loop so that the north-seeking pole of the permanent magnet and the two south-seeking poles will be opposite. The like poles will oppose each other, and the loop will continue to rotate in a counterclockwise direction.

**Q33. If a current-carrying conductor is placed in a magnetic field so that the current appears to be flowing into the page and the polarity of the permanent magnet is such that the north-seeking pole is on your right, what will be the direction of force on the electron stream?**

**Your Answer Should Be:**

**A33.** The direction of force on the electron flow would be **downward.**

## DC Motors and Generators

In electric motors many loops of wire are wound around a core. This assembly is called an **armature.** Each loop is connected to a commutator segment that makes contact with the carbon brushes as the armature rotates. The use of many loops provides smoother operation and considerably more power than a single loop. An end view of an armature is shown in the figure below.

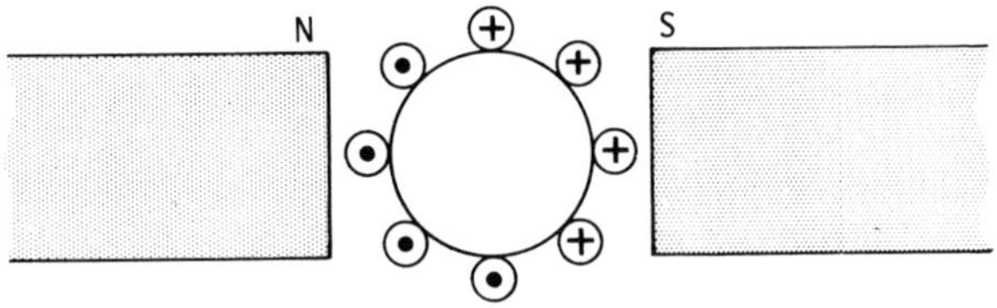


Large electric motors use electromagnets in place of permanent magnets. It is possible to obtain a stronger magnet for the same physical size by using the electromagnet.

If current flowing through a wire creates a magnetic field, it seems only reasonable that a wire moving through a magnetic field causes a current flow. The DC generator operates by use of this principle. An armature (similar to one in an electric motor) is rotated in a magnetic field. The turns of wire cut the lines of force, and a current is caused to flow in the wire loops of the armature. Connections to the commutator provide an electric current output.

Another left-hand rule is used to determine the direction of the induced electromotive force in a generator. Place the thumb, index finger, and middle finger of your left hand so that they are perpendicular and at right angles to each other. Point the thumb in the direction of motion (rotation) of the conductor (armature) and point the index finger in the direction of the magnetic flux. The middle finger will then indicate the direction of the induced current (electron flow).

This procedure is shown in the figure below.

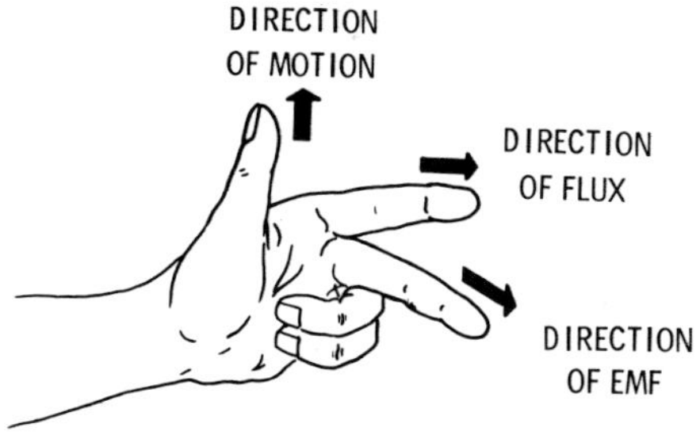


**Left-hand rule for generators.**

## Meters

Both permanent magnets and electromagnets are used in meter movements. Their operation is similar to that of the electric motor previously described. In the meter movement the electromagnet does not rotate through 360° as it did in the motor; instead, it rotates through an arc of approximately 150°.

The electromagnet is positioned between the poles of a permanent horseshoe magnet, as shown in the figure below.

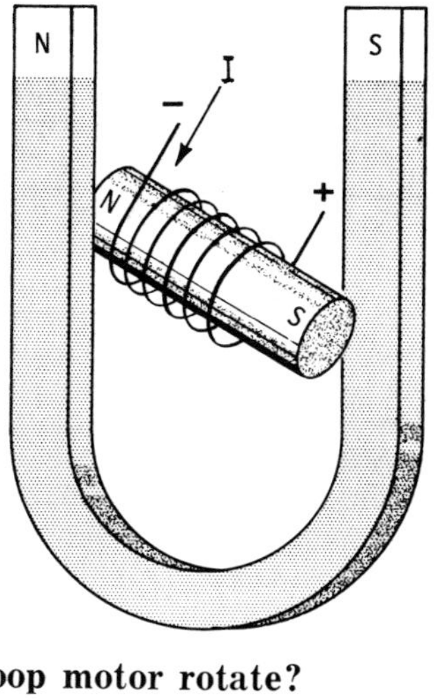


**Meter principle.**

**Q34. How long will a single-loop motor rotate?**

**Q35. If the magnets in a generator are placed so that the north-seeking pole is on the right and the motion of the conductor is downward, what will be the direction of the induced current flow?**

**Your Answers Should Be:**

**A34.** A single-loop motor will rotate **as long as current flows through the loop.**

**A35.** The direction of the induced current flow is **out of the page** (left-hand rule for generators).

When no current is flowing through the coil, a spring holds the coil in the position shown. When current does flow, a magnetic field is developed around the coil with the north-seeking pole on the left and the south-seeking pole on the right in the figure above (left-hand rule for coils). Thus, the two north-seeking poles and the two south-seeking poles are opposite each other. The like poles repel and the coil rotates. How far the coil rotates depends on the strength of the electromagnet and its ability to overcome the tension applied by the spring. A pointer connected to the moving coil moves across a calibrated scale, making it possible to use the meter as a measuring device.

## WHAT YOU HAVE LEARNED

1. Magnetism is a property of certain materials to attract and repel each other.
2. Magnetized materials have north (north-seeking) and south (south-seeking) poles. Magnetic force lines flow from south to north inside the material and north to south outside.
3. Some materials may be magnetized by stroking them with a magnet or by passing direct current through a coil wrapped around them.
4. A permanent magnet has a high retentivity (retains its magnetism). A temporary magnet has low retentivity.
5. Permanent magnets should be stored in a manner which permits the external field to be concentrated in a path of low flux opposition. Bar magnets are stored with N and S poles adjacent. A keeper is placed across the poles of a horseshoe magnet.
6. Reluctance is the opposition offered to the flow of magnetic flux lines. Air has a higher reluctance than iron.

7. Number of flux lines (maxwells) is directly proportional to the magnetomotive force exerted and indirectly proportional to the reluctance of the material through which the flux lines pass.
8. Permeability of a material is a measure of its ability to be magnetized. Low reluctance indicates high permeability.
9. An electromagnet is a device that has been or is being magnetized electrically.
10. Current through a conductor generates a magnetic field. If the thumb of the left hand points in the direction of electron flow, the fingers curl in the direction taken by the flux lines.
11. A coil of wire develops a stronger magnetic field than a single conductor. Field strength is directly proportional to ampere-turns (number of coil turns and the amount of current flowing). Field strength is indirectly proportional to the diameter of the coil.
12. Magnetomotive force of a coil can be determined by multiplying ampere-turns by a reluctance constant. Reluctance for air is 1.257.
13. Residual magnetism is the amount of magnetism remaining in an electromagnet after current flow has ceased.
14. Hysteresis (difficulty in realigning magnetic direction of molecules) causes changes in the magnetic field to lag changes in the current.
15. Each magnetic material has a limit to which it can be magnetized. The limit of magnetic strength is called saturation, and is the point at which a maximum number of molecules have been aligned in the same magnetic direction.
16. Solenoids are electromagnets with movable iron cores.
17. Electromagnets are used in motors, generators, meters, and other devices that make use of the electrical effects of a magnetic field.
18. The right-hand rule for motors states: With thumb, index finger, and middle finger of the right hand at right angles to each other, the index finger pointing in

the direction of magnetic flux, and the middle finger pointing in the direction to current flow, the thumb will indicate the direction in which the conductor will move.

19. The left-hand rule for generators uses the same principle: If the left thumb points in the direction of the conductor movement, and the index finger points in the direction of the magnetic flux, the middle finger will indicate the direction of current flow in the conductor.

# 7

# What Is Alternating Current?

## *What You Will Learn*

When you have finished this chapter you will be able to explain what alternating current is. You will not only know how AC currents are generated, but you will be able to recognize AC and pulse forms. In addition, you will be able to describe the different types of AC waveforms.

Alternating current (AC) differs from direct current (DC). The electrons in a direct current always flow in the same direction. Electrons in an alternating current reverse directions periodically.

Electrons always flow in the same direction in a DC circuit.

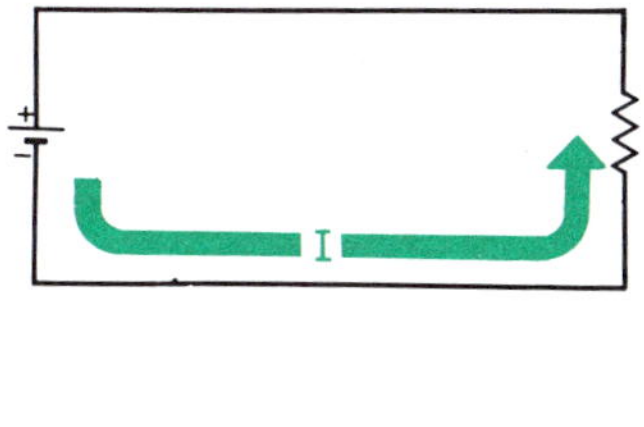


Electrons reverse direction regularly in an AC circuit.

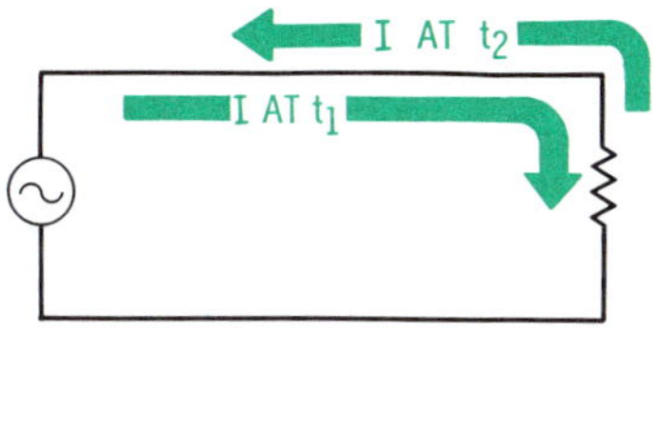

## ALTERNATING CURRENT SOURCES

Alternating current is the most common type of electricity used. Because its electron flow reverses direction rapidly, the value of an alternating voltage is easily increased or decreased by passing it through a transformer. For example, when it is necessary to transmit electricity over great distances (from the power plant to a city miles away), the power-plant voltage is increased by a transformer to a very high value (such as 69,000 volts) and sent through a relatively small-diameter transmission line. If the AC electricity were sent at the power-plant voltage (such as 4,160 volts), the diameter of the transmission line would have to be much larger. This is because a lower voltage makes a higher current flow necessary. Thus, a larger-diameter line from the power plant to the user would be required. At the city end of the transmission lines the voltage is reduced by transformers, and then still further reduced to 240 and 120 volts by another transformer at the utility pole outside the user's house. It is far more difficult and costly to change DC voltages than to change AC voltages.

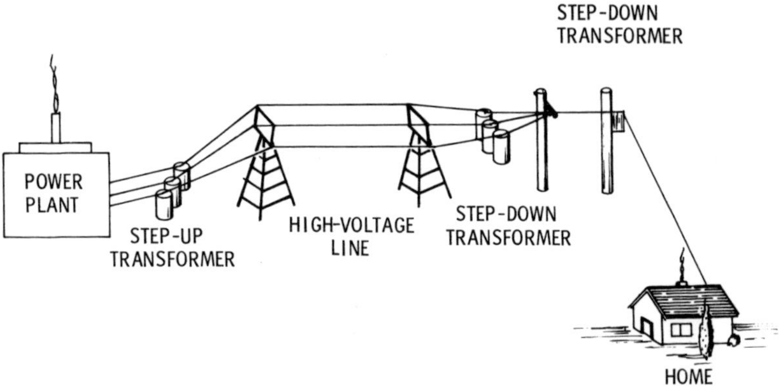


**Alternating current can be transmitted long distances at low cost.**

Alternating currents are generated in very large quantities in power plants. The electricity used in your home comes from a power plant. The total current capacity of a power plant can reach several thousand amperes at 4,160 volts.

Another source of AC, usually for specialized purposes, is the **oscillator.** The current output of an oscillator is quite small, usually in the milliampere range. Oscillators are used primarily as signal sources in electronic equipment.

## ALTERNATING-CURRENT APPLICATIONS

Large amounts of alternating current are used in homes for illumination, heating, cooking, and the operation of appliances. In industry, alternating current is used to operate motors and for many other applications.

Most of the alternating current used in homes and industry is produced by generators in large power plants. In most cases these generators are driven by turbines powered by either steam or falling water.

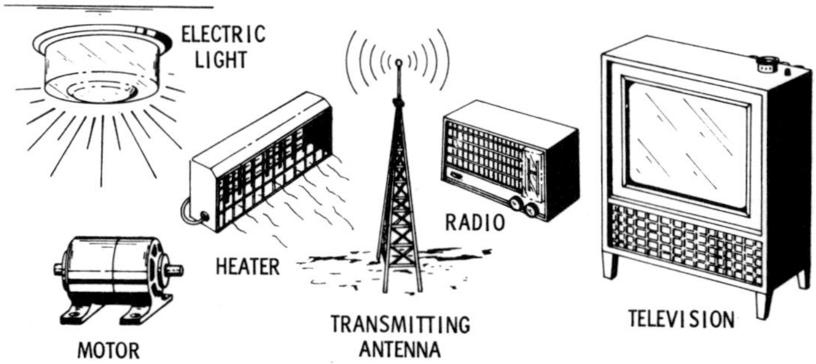


**AC is used to power many familiar devices.**

**Q1. The electricity usually used in homes is not DC current but -- current.**

**Q2. The electricity used in homes is produced by generators in power plants. Power plants are one important ------ of AC.**

**Q3. Name the most common kind of current used for lighting.**

**Q4. It is difficult and expensive to transmit DC current over long distances, or to raise or lower DC voltage. State two advantages of alternating current.**

**Q5. Alternating current is used in many applications in homes and in industry. Two sources of power used to generate this AC are ----- and ------------.**

**Your Answers Should Be:**

**A1.** The electricity usually used in homes is not DC current but **AC** current.

**A2.** The electricity used in homes is produced by generators in power plants. Power plants are one important **source** of AC.

**A3.** The most common kind of current used for lighting is **AC.**

**A4.** Two advantages of alternating current are: **It is less expensive and easier to transmit over long distances,** and **it is easier to obtain and keep a constant voltage level.**

**A5.** Alternating current is used in many applications in homes and industry. Two sources of power used to generate this AC are **steam** and **falling water.**

## WAVEFORMS

**Waveforms** are pictures showing how currents and voltages change over a period of time. The value of voltage or current is usually represented in the vertical direction, while time is represented in the horizontal direction. For example, the illustration below shows that most DC waveforms are straight lines.

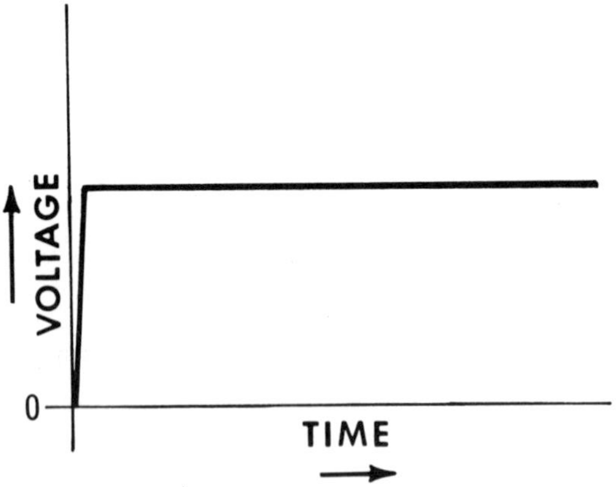


DC waveforms are usually a straight line.

**Pulsating** DC voltages may have various shapes, but the two most common types are shown here.

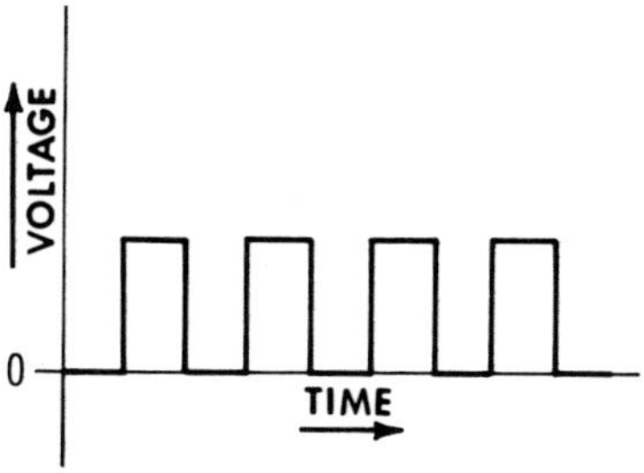


Pulsating DC voltage.

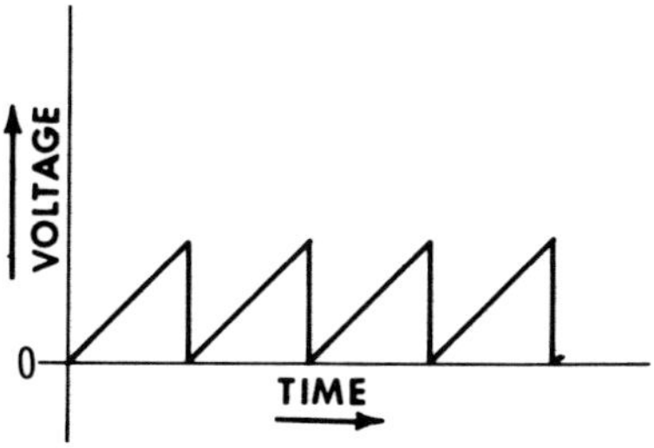


Sawtooth DC voltage.

By far the most common AC waveform is the **sine wave.** In fact, the sine wave is so widely used that when we think of AC, we automatically think of sine waves.

Household AC is a sine wave.

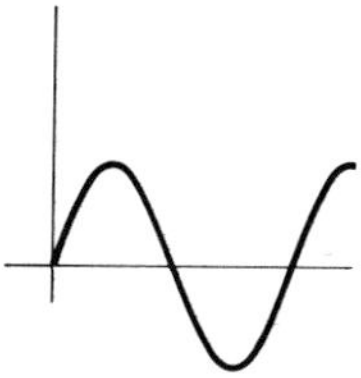

**Q6. A waveform is a picture of how currents or voltages change over a period of time. DC waveforms usually look like which of the following?**

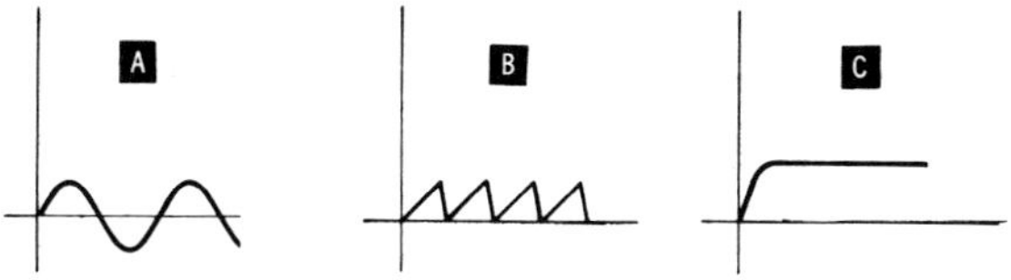


**Q7. AC currents and voltages usually change regularly in a smoothly curved form. Which of the following is the usual AC waveform?**

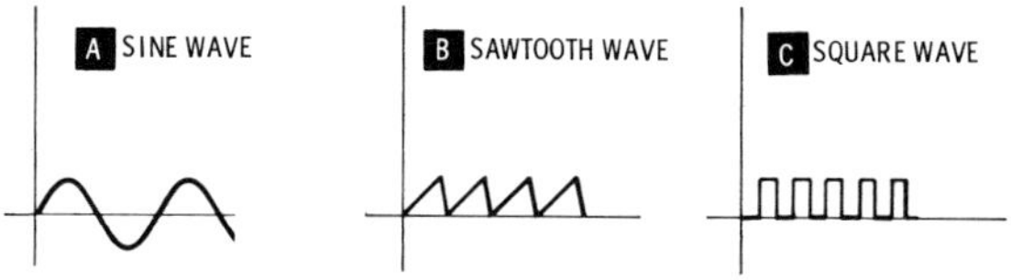


**Q8. The most common DC waveform is a _____________. The most common AC waveform is a ________.**

**Your Answers Should Be:**

**A6.** The usual DC waveform is **(c),** a straight line.

**A7.** The usual AC waveform is **(a),** a sine wave.

**A8.** The most common DC waveform is a **straight line.** The most common AC waveform is a **sine wave.**

## GENERATION OF A SINE WAVE

A sine wave is the most common AC waveform. It is also the simplest. You can visualize the way it is generated by looking at this illustration.

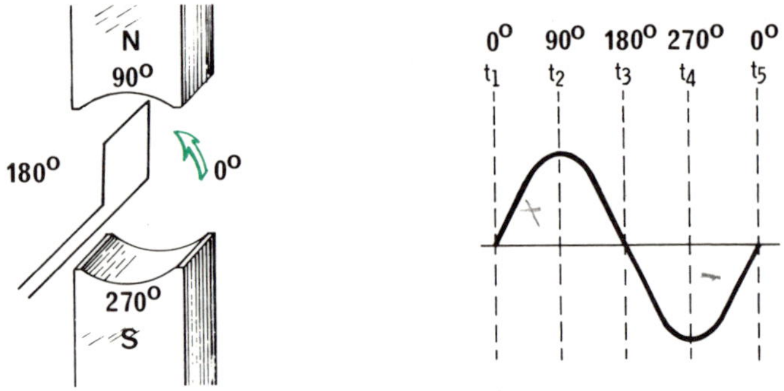


**A rotating coil can create a sine wave.**

As the coil cuts the lines of force between the magnetic poles, a voltage is produced that will cause current to flow if the coil is connected to a complete circuit. As the coil rotates at a constant speed, it cuts more and more lines of force in the magnetic field, and the voltage increases. At 90° it is moving at right angles to the lines of force, so it cuts the maximum number of lines per second. Therefore voltage is maximum.

At 180° (and at 0°) the coil is moving parallel to the lines of force and, therefore, cutting none. Thus, the voltage generated is zero. Beyond 180° the coil is cutting lines of force in the opposite direction, so the generated voltage has the opposite polarity—negative in this case. The output waveform of the generator is a sine wave like the one shown.

Imagine a line like the hand of a clock. This line is called a **voltage vector.** A voltage vector rotates counterclockwise through the full 360° of a circle. The distance measured from the end of the voltage vector to the base line at any time during the rotation of the vector represents the exact value of voltage at that instant. As you can see, the value of voltage is zero at 0° and 180°. At 90° the value of voltage is maximum positive, and at 270° it is maximum negative.

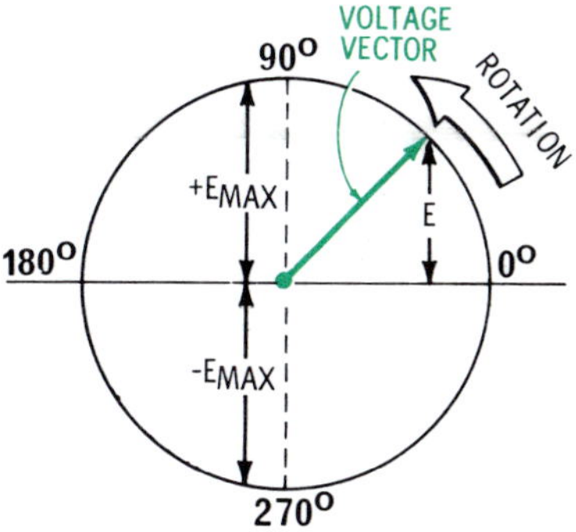


**The rotating voltage vector creates a sine wave.**

One complete cycle of a sine wave (from zero to a positive peak, down to a negative peak, and back to zero) simply represents one complete rotation of the voltage vector.

The simple sine wave is the building block from which all AC waveforms are constructed. Even sawtooth and square waves are really just complicated combinations of simple sine waves.

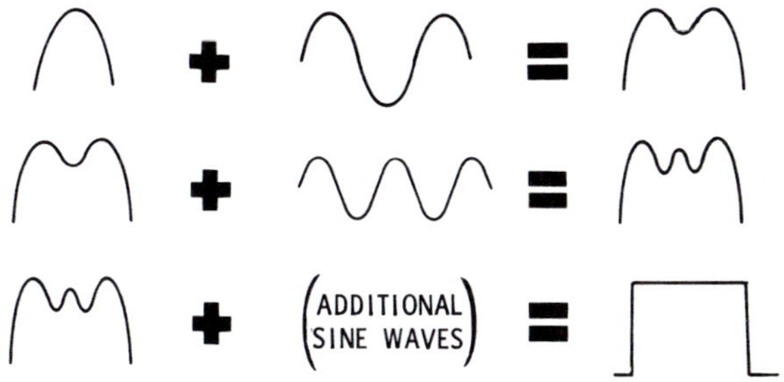


**A square pulse is a combination of many sine waves.**

**Q9. A sine wave represents the rotation of a line called a voltage ______.**

**Q10. With the right mixtures of sine waves, can sawtooth and square waves be made?**

**Your Answers Should Be:**

**A9.** A sine wave represents the rotation of a line called a voltage **vector.**

**A10. Any kind of AC waveform can be made with the right mixture of sine waves.**

## SINE-WAVE MEASUREMENT

In a previous chapter you learned that DC voltage has only one value. This value is measured in volts. In AC, however, the voltage is constantly changing, so no one voltage value exists for more than an instant. Looking at a sine wave you can see that it reaches a certain peak. That value is known as maximum voltage, or peak voltage ($E_{max}$). Notice that the waveform has the same shape and value both above and below the zero line.

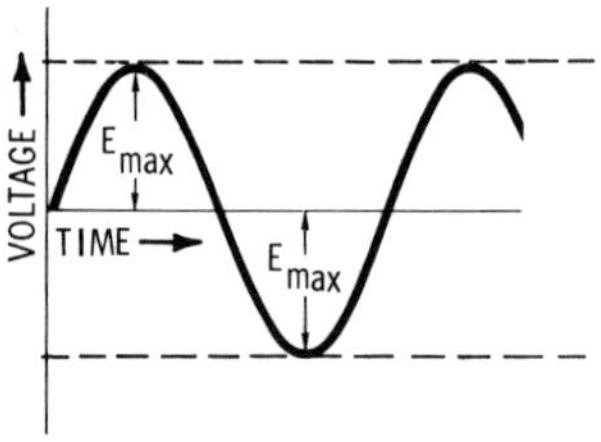


**A sine-wave voltage reaches a certain peak value.**

A more practical value of AC voltage and current is the **rms** value (rms stands for root-mean-square). The rms value is the actual "working value" of a voltage or current and is equivalent to the DC value which would accomplish the same amount of work.

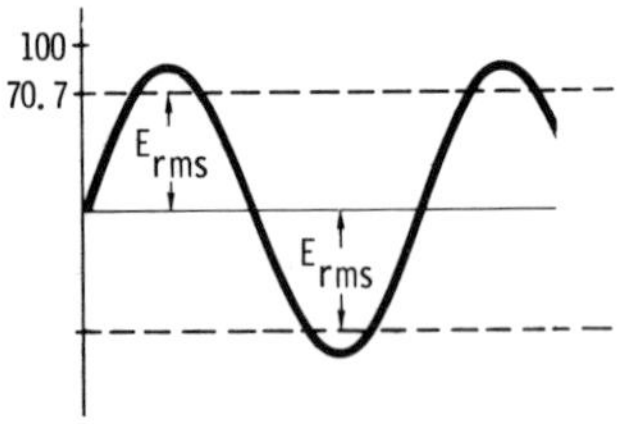


**Rms voltage is the working voltage.**

The current and voltage value most often used is rms. In fact, the standard household voltage of 120 volts is the rms value. An rms voltage of 120 volts corresponds to a peak voltage of 170 volts. In all sine waves the rms value is equal to 0.707 of the peak value. Conversely, the peak value is equal to 1.41 times the rms value.

You have seen how AC voltage first increases in value to a peak value, then decreases to zero, increases to a negative peak value, and then returns to zero. This sequence is known as a **cycle.** The cycle is normally repeated many times each second and is identical in waveform with the one before it and the one following it.

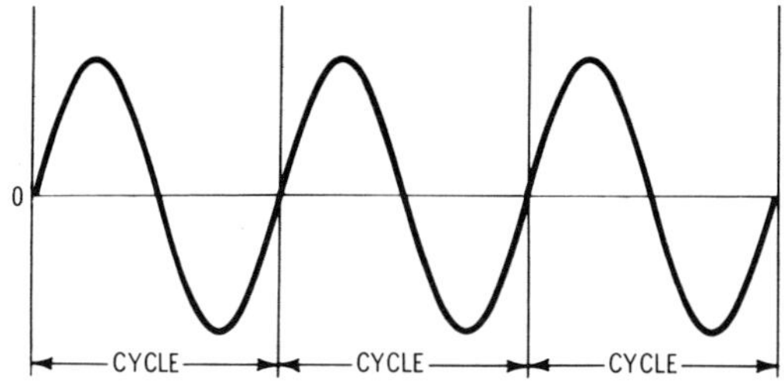


**All sine-wave cycles are normally identical.**

**Q11. Another name for peak voltage is _______ _______.**

**Q12. Rms is the working value of voltage. Which is less, rms or peak voltage?**

**Q13. Maximum voltage is also called ____ _______.**

**Q14. Working voltage is called ___ _______.**

**Q15. The symbol for maximum voltage is $E_{max}$. The symbol for rms voltage is ____.**

**Q16. Rms voltage is 0.707 times peak voltage. If peak voltage is 100, rms voltage is ____.**

**Q17. Peak voltage is 1.41 times rms voltage. That is, if rms voltage is 100, peak voltage is ____.**

**Q18. To change rms voltage to peak voltage (which is larger), multiply by ____. To change peak voltage to rms voltage, multiply by ____.**

**Q19. To give a complete picture of how an AC voltage (or current) changes, a waveform diagram must show at least ___ complete cycle(s).**

**Your Answers Should Be:**

**A11.** Another name for peak voltage is **maximum voltage.**

**A12.** Rms is the working value of voltage. **Rms is less than peak voltage.**

**A13.** Maximum voltage is also called **peak voltage.**

**A14.** Working voltage is called **rms voltage.**

**A15.** The symbol for maximum voltage is $E_{max}$. The symbol for rms voltage is $\mathbf{E_{rms}}$.

**A16.** Rms voltage is 0.707 times peak voltage. If peak voltage is 100, rms voltage is **70.7.**

**A17.** Peak voltage is 1.41 times rms voltage. That is, if rms voltage is 100, peak voltage is **141.**

**A18.** To change rms voltage to peak voltage (which is larger), multiply by **1.41.** To change peak voltage to rms voltage, multiply by **0.707.**

**A19.** To give a complete picture of how an AC voltage (or current) changes, a waveform diagram must show at least **one** complete cycle.

## Frequency

It is sometimes necessary to know how many times a cycle is repeated each second. The number of cycles completed each second by a given AC voltage is called the **frequency.**

***FREQUENCY IS THE NUMBER OF CYCLES PER SECOND***

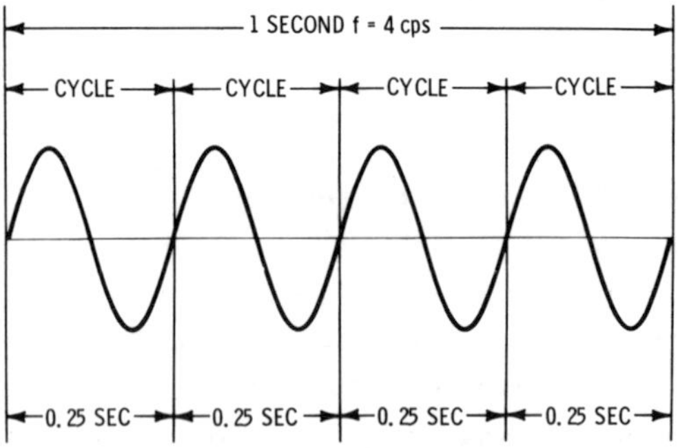


***PERIOD IS THE TIME REQUIRED TO COMPLETE ONE CYCLE***

Frequency is measured in **cycles per second** (cps). The symbol for frequency is **f.** The most common type of current (household current) has a frequency of 60 cps in most localities.

The time required for one cycle is called a **period.** A period is measured in **seconds, milliseconds,** or **microseconds.** A millisecond is 1/1,000 of a second, and a microsecond is 1/1,000,000 of a second. Since household electricity has a frequency of 60 cps, its period is 1/60 of a second, or 0.0167 second. This time can also be expressed as 16.7 milliseconds.

The period of a sine wave represents the time needed for the voltage vector to make one complete rotation. The frequency of a sine wave depends on how rapidly the voltage vector rotates.

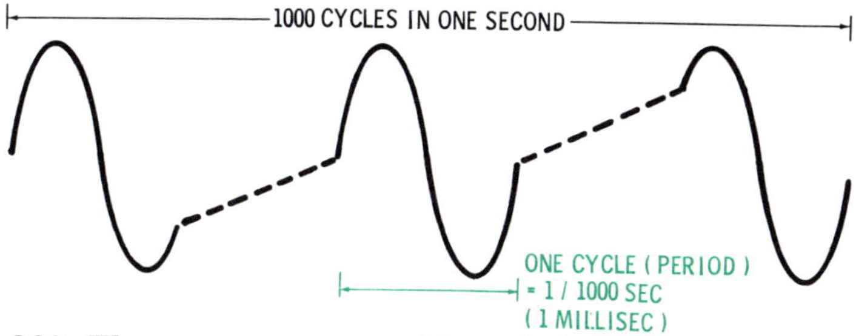


**Q20. The ---------- of household current is 60 cps.**

**Q21. Frequency is usually measured in ------ per second.**

**Q22. If the frequency of a voltage is 60 cps, how would you find its period?**

**Q23. How many milliseconds are there in a second? How many microseconds are there in a millisecond?**

**Q24. If the frequency of an AC voltage is 1 million cycles per second, state its period in the most convenient unit.**

**Q25. In a certain sine wave the rms voltage is 20 volts. What is the peak voltage?**

**Q26. A sine wave has a peak value of 70.7 volts. What is the effective (rms) voltage?**

**Q27. If a current has a frequency of 100 cps, what is the period of one cycle? Give two answers—one in seconds and one in milliseconds.**

**Your Answers Should Be:**

**A20.** The **frequency** of household current is 60 cps.

**A21.** Frequency is usually measured in **cycles** per second.

**A22.** **Divide 1 second by 60 cycles.** Thus the period of a 60-cps current is 1/60 of a second, or 0.0167 second.

**A23.** There are **1,000 milliseconds in a second,** and **1,000 microseconds in a millisecond.**

**A24.** **1 microsecond.**

**A25.** $20V_{rms} \times 1.41 = 28.2V_{max}$

**A26.** $70.7V_{max} \times 0.707 = 50V_{rms}$

**A27.** **0.01 second, or 10 milliseconds.**

## PULSES

You have been introduced to the definition and methods of pulse generation in DC electricity. In modern electronics, considerable use is made of these pulses. In the following paragraphs you will see two practical applications. Later, you will learn more about pulse behavior in electrical circuits.

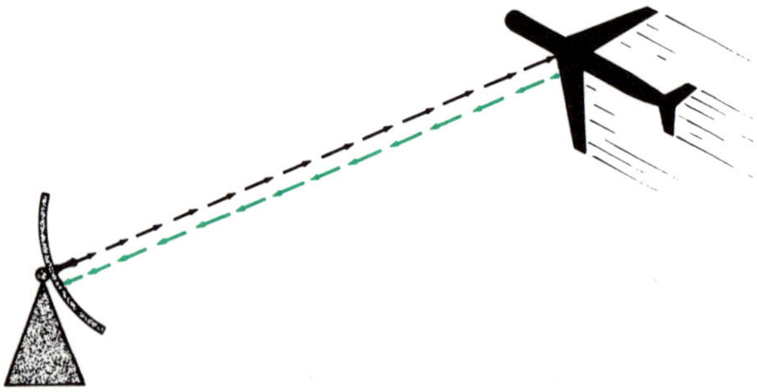

**Radar bounces pulses of energy off distant objects.**

A radar set sends a pulse of energy into space. This pulse travels at the speed of light. The pulse hits an object (an airplane, for instance) and is reflected back to the set. By measuring the time it takes for this signal to travel to the object and return, the distance between the set and the object can be calculated.

Digital computers are often used in calculating difficult problems. They are the so-called electronic brains. A very simple digital principle is an assembly line; for example, a series of cans moving on a conveyor belt. A light beam to a photoelectric cell is cut every time a can goes by, and the resulting signal is then counted. Computers employ various operations with pulses at very high speeds.

**Q28. Look at the sine wave in the figure below and write down the values for:**

**$E_{rms}$ = ____ Period in seconds = ____ ---**

**$E_{max}$ = ____ Frequency = ____ ---**

**Period = __ ------------**

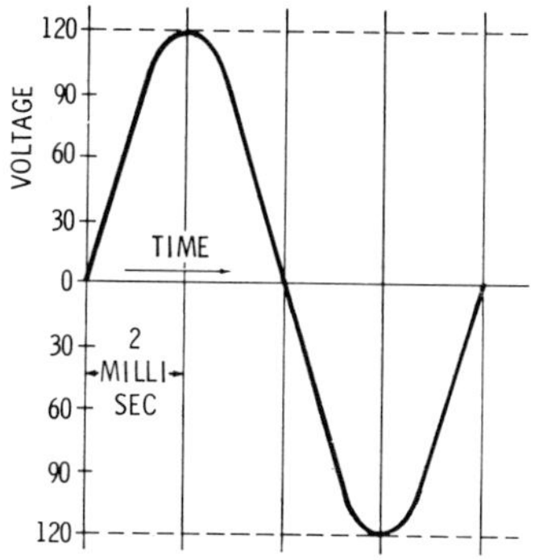


**Q29. Which of the following waveforms would you especially expect to find in a digital computer?**

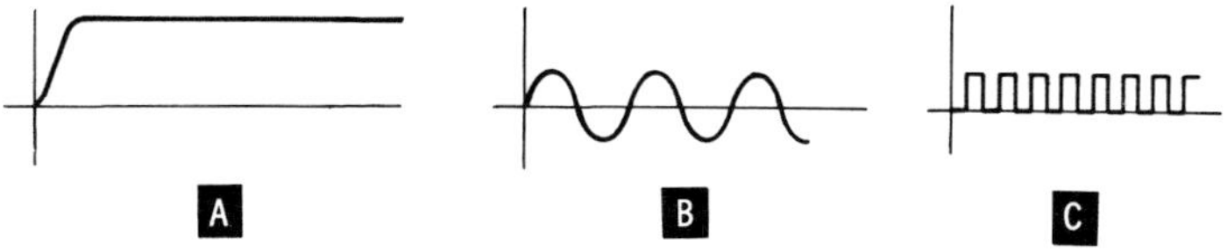


**Q30. The signal from a radar set looks like this:**

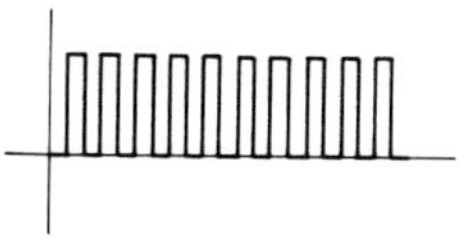

**What type of signal does a radar set send out?**

**Q31. Name one characteristic that a radar set and a digital computer have in common.**

**Your Answers Should Be:**

**A28.** $E_{rms} = 84.8V$ Period = **0.008 sec**

$E_{max} = 120V$ Frequency = **125 cps**

Period = 8 **milliseconds**

**A29.** **(c)** is a pulse waveform which would be found in a digital computer. The other two might be found in the operation of some of its circuits.

**A30.** A **pulse** signal is the type of signal a radar set sends out.

**A31.** One characteristic that a radar set and a digital computer have in common is that both use **pulse waveforms.**

## SAWTOOTH VOLTAGE

Most of the pulses described in the last section are of the regular square-wave type. Another very important type of pulse is the sawtooth waveform.

Sawtooth waves are primarily used in accurately timing the rate of sending square-wave pulses. In other words, sawtooth waveforms are used when time measurement is required, and they act as triggering devices.

Some of the most important uses of sawtooth pulses are in television sets and oscilloscopes, and for timing radar pulses.

## PULSE MEASUREMENT

Pulses are usually described in terms of four parts—**base line, leading edge, peak,** and **trailing edge.** It is easiest to understand these parts by looking at them.

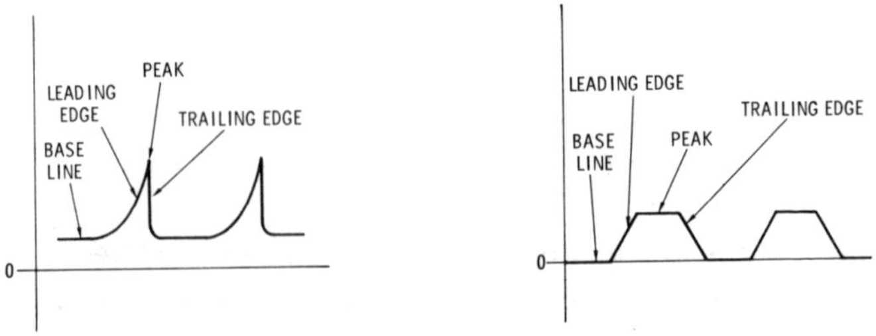


**The parts of a pulse waveform.**

It is often necessary to know how long it takes a pulse to rise from the base line to its peak **(rise time)** or to go from its peak value back to the base line **(decay time).** These times are measured between 10% and 90% of peak value.

**Rise time and decay time are measured between 10% of peak and 90% of peak.**

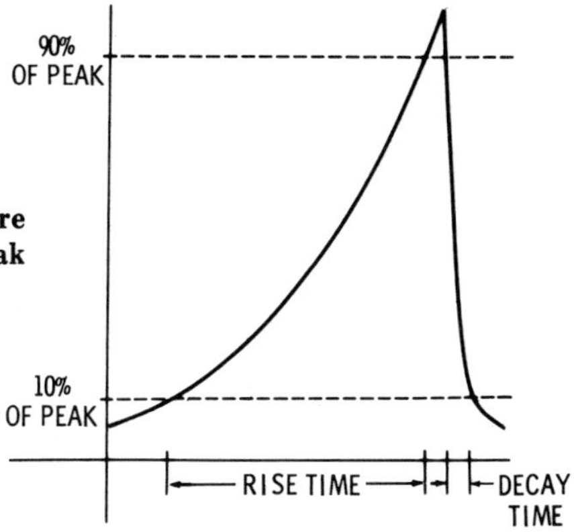


Like AC sine waves, pulses have an rms value. Unfortunately, there is no simple formula to find the rms value of a pulse. The rms value of a pulse is its working value.

When pulses are repeated at a regular rate, the number of pulses per second is called the **repetition rate.**

**Q32. Radar, television, and oscilloscopes use special pulses that look like the teeth of a saw. This kind of pulse is called a ________ voltage.**

**Q33. Radar uses both ordinary pulses and ________ ________.**

**Q34. Name three types of equipment which use sawtooth voltages.**

**Q35. Rise time is the time during which the pulse goes from 10% of peak to ____ of peak. During decay the pulse decreases from ____ of peak to ____.**

**Q36. The time required for a pulse to go from 10% to 90% of peak is called ____ time. The same time along the trailing edge is called _____ time.**

**Q37. What is the working value of a pulse called?**

**Q38. The number of times a pulse is repeated per second is called its __________ ____.**

**Your Answers Should Be:**

**A32.** Radar, television, and oscilloscopes use special pulses that look like the teeth of a saw. This kind of pulse is called a **sawtooth** voltage.

**A33.** Radar uses both ordinary pulses and **sawtooth voltages.**

**A34.** Three types of equipment which use sawtooth voltages are **television, radar, and oscilloscopes.**

**A35.** Rise time is the time during which the pulse goes from 10% of peak to **90%** of peak. During decay the pulse decreases from **90%** of peak to **10%.**

**A36.** The time required for a pulse to go from 10% of peak to 90% of peak is called **rise** time. The same time along the trailing edge is called **decay** time.

**A37.** The working value of a pulse is its **rms value.**

**A38.** The number of times a pulse is repeated per second is called its **repetition rate.**

## WHAT YOU HAVE LEARNED

1. AC current changes direction regularly.
2. Most household appliances and electronic devices use AC electricity.
3. AC waveforms are usually sine waves.
4. AC is created by generators and oscillators.
5. The working voltage of a sine wave AC is the rms voltage, which is 0.707 times the peak voltage.
6. A cycle is one complete change from zero to the positive peak value, back through zero to the negative peak, and back to zero. This represents the rotation of a voltage vector around the 360° of a circle.
7. Frequency is the number of cycles generated in a second, and a period is the time it takes to complete one cycle.
8. The four main parts of a pulse waveform are the base line, the leading edge, the peak, and the trailing edge.

# 8

# Calculating Resistance

## *What You Will Learn*

**In this chapter you will learn how to draw the schematic of a basic AC circuit. You will determine when and how Ohm's law and the power formulas can be used in AC circuits. You will learn how to simplify combinations of resistances in an AC circuit to find the equivalent resistance. You will be able to tell when voltage and current are in phase. You will learn about skin effect and where and when it is found.**

### BASIC AC CIRCUIT

The basic AC circuit is very similar to the basic DC circuit. The only difference is that an AC generator is used as a voltage source instead of a battery.

**AC generator symbol.**

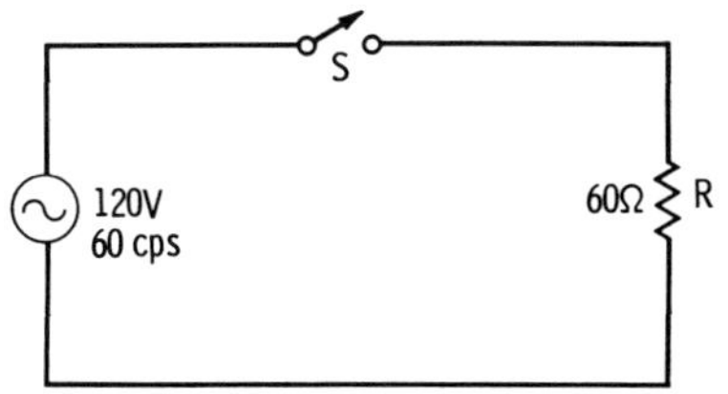


**A basic AC circuit.**

In any AC circuit the voltage shown at the generator is the rms voltage. The frequency shown is in cps (cycles per second), and the resistance is in ohms.

## OHM'S LAW

Ohm's law, as you learned it for DC, can also be used when AC is supplied to a resistive circuit. As stated previously, the rms voltage is the working voltage of AC electricity. In a basic AC circuit, all calculations will be made with rms voltage and current values, unless other values are specified. Ohm's law also applies to peak values, but peak values are not generally very useful.

Let us look at a basic AC circuit and see what happens as the switch is closed.

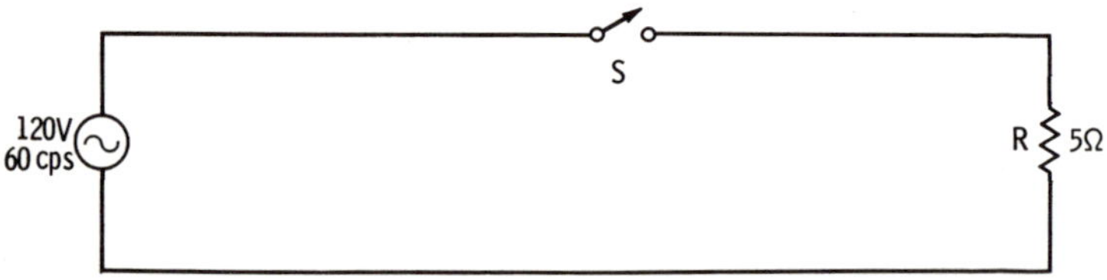


**The values you calculate are working values.**

From Ohm's law we find that the current in the above circuit will be $I = \frac{120}{5} = 24$ amperes. But, since this is an AC circuit, we also know that both the 120 volts and the 24 amperes are rms values; that is, they are the working values. Since AC voltage and currents appear as sine waves, the voltage actually varies. The current also varies, but its numerical value is always one-fifth that of the voltage.

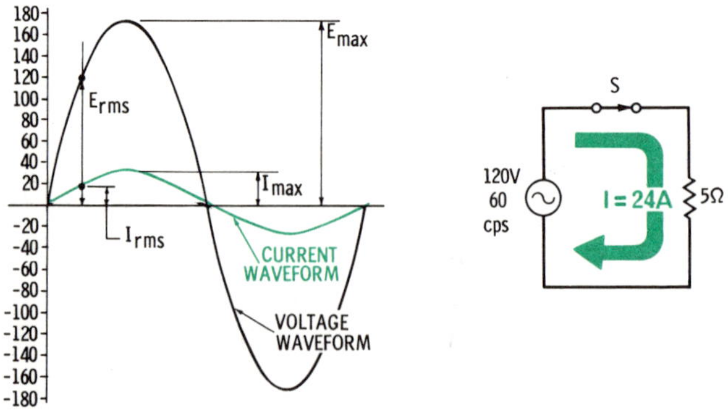


**Voltage and current in a basic AC circuit.**

## PHASE

When both the voltage and the current rise and fall together in exactly the same fashion, they are said to be **in phase.** When they do not, they are **out of phase.**

When an AC circuit contains only pure resistance, the voltages and currents will always be **in phase.** When voltages and currents in an AC circuit are in phase, Ohm's law can be applied in the same manner as in DC circuits, provided you use the same kind of values (rms, etc.) for the voltages and currents.

**Q1. A basic AC circuit consists of a ----------, -----------, ------, and ---- (----------).**

**Q2. Draw a basic AC circuit in which the generator has an rms voltage of 12 volts, a frequency of 60 cps, and the load is 10 ohms. Check your circuit against the correct one on the next page.**

**Q3. Find the current in this AC circuit, using Ohm's law.**

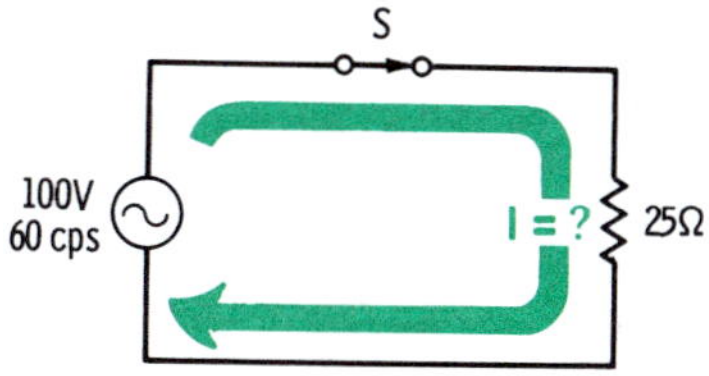


**Q4. Ohm's law applies to any AC circuit that contains only -----------.**

**Q5. When using Ohm's law with AC circuits, you will usually use the working value of voltage, which is ----.**

**Q6. When current and voltage are zero at the same time, maximum at the same time, and vary in the same fashion, they are -- -----.**

**Q7. When voltage and current are in phase, they (have the same maximum value, reach zero at the same time).**

**Q8. Voltage and current are always in phase in AC circuits containing only -----------.**

**Your Answers Should Be:**

A1. A basic AC circuit consists of a **generator, conductors, switch,** and **load (resistance).**

A2. Your basic AC circuit should look like this:

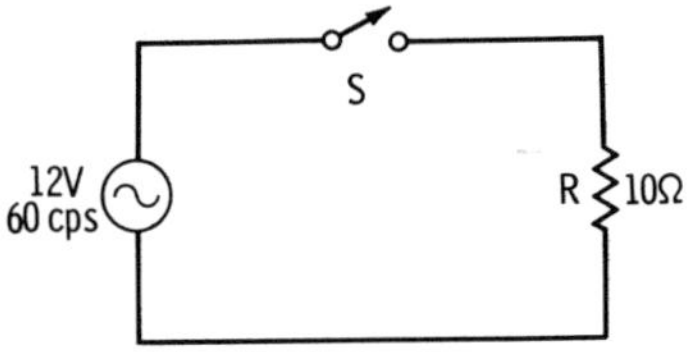


A3. Ohm's law is: $E = I \times R$, or $I = \frac{E}{R}$

In this case, $I = \frac{100}{25} =$ **4 amperes**

A4. Ohm's law applies to any AC circuit that contains only **resistance.**

A5. When using Ohm's law with AC circuits, you will usually use the working value of voltage, which is $\mathbf{E}_{rms}$.

A6. When current and voltage are zero at the same time, maximum at the same time, and vary in the same fashion, they are **in phase.**

A7. When voltage and current are in phase, they **reach zero at the same time.**

A8. Voltage and current are always in phase in AC circuits containing only **resistance.**

## POWER IN A BASIC AC CIRCUIT

As you learned in the study of DC circuits, power is the work done by the current, and it is measured in watts. In DC circuits you have seen how power equals the voltage multiplied by the current. You also learned that the power in a resistance equals the value of the resistance multiplied by the current squared.

Voltage and current must be in phase if they are used together in an AC formula. For example, $P = E \times I$ is true only when voltage and current are in phase.

It is therefore a good idea when working with AC circuits to use only $P = I^2R$ for finding power. When using this formula, it makes no difference whether voltage and current are in phase or out of phase.

## AC CIRCUITS WITH RESISTANCES IN SERIES

This is an AC circuit with resistances in series.

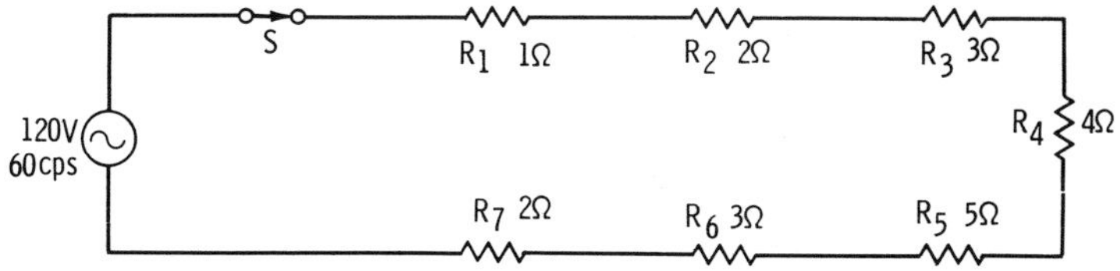


**An AC circuit with resistances in series.**

To find the current in this circuit, you must find the equivalent resistance of the resistors in series (the one resistance that can replace all the other resistances).

This AC series circuit can be treated in the same way as a DC series circuit. Since the resistors are all in series, the equivalent resistance is the sum of all the resistances.

$$R_{eq} = R_1 + R_2 + R_3 + R_4 + R_5 + R_6 + R_7$$

**Q9. Which power formula can only be used when voltage and current are in phase?**

**Q10. Are voltage and current in phase in the basic AC circuit above? Explain how you know.**

**Q11. Which formula or formulas can be used to find power in the basic AC circuit above?**

**Q12. Write the power formula that can be used when voltage and current in an AC circuit are out of phase.**

**Q13. What is the equivalent resistance in the following circuit?**

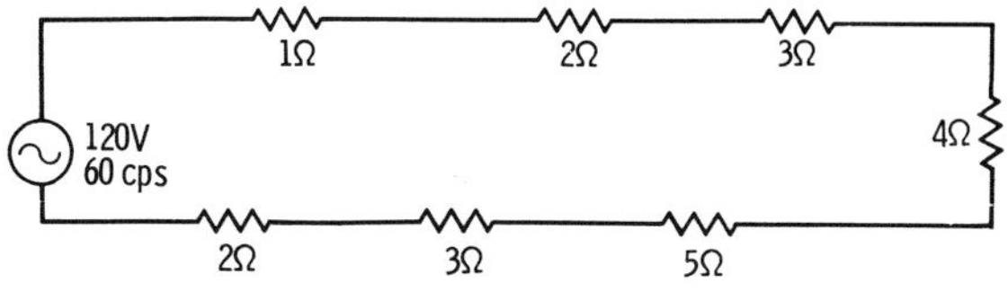

**Your Answers Should Be:**

A9. **P = I × E** can only be used when voltage and current are in phase.

A10. **Yes,** because voltage and current are always in phase in a circuit that contains only resistance.

A11. **$P = I^2R$. P = I × E.**

A12. **$P = I^2R$.**

A13. **20 ohms.**

## AC CIRCUITS WITH RESISTANCES IN PARALLEL

This is an AC circuit with resistances in parallel.

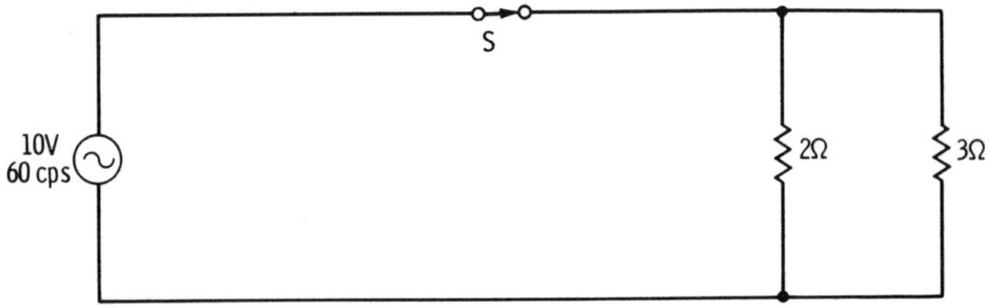


The formula for the equivalent resistance of two resistors in parallel is:

$$R_{eq} = \frac{R_1 \times R_2}{R_1 + R_2} = \frac{2 \times 3}{2 + 3} = \frac{6}{5} = 1.2 \text{ ohms}$$

Here is another AC circuit with two **equal** resistances in parallel.

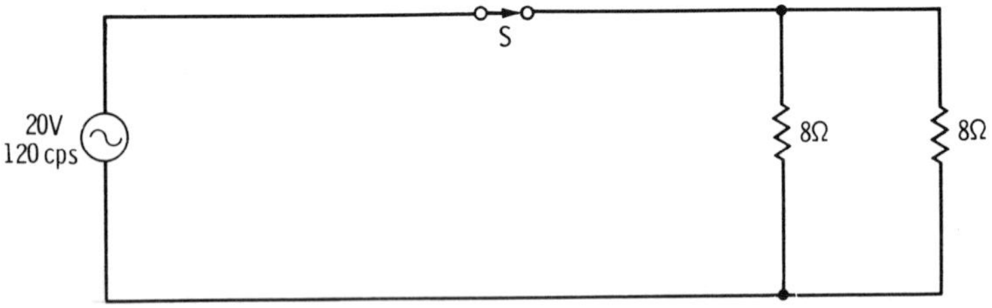


The formula for the equivalent resistance of two **equal** resistors in parallel is:

$$R_{eq} = \frac{R_1}{2} = \frac{8}{2} = 4 \text{ ohms}$$

Did you notice these three important facts about parallel resistances?

1. The equivalent resistance is always smaller than either of the parallel resistances.
2. When two parallel resistances are equal, the equivalent resistance is one half as large as either resistance.
3. Frequency does not enter into the calculations.

Here is an example of how to simplify a combination of parallel resistances.

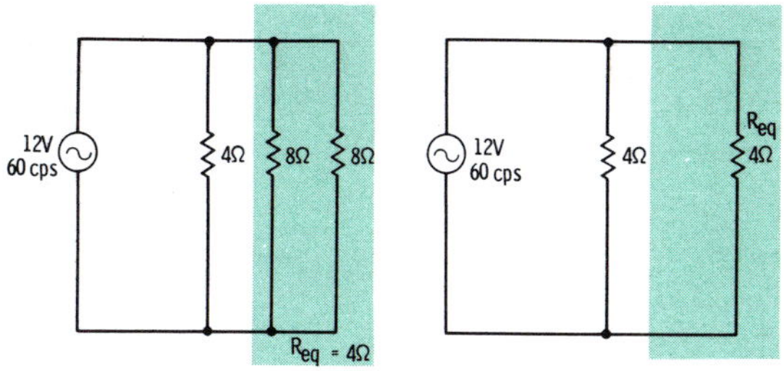


**Combinations of resistances can be simplified one step at a time.**

**Q14.** **Can Ohm's law be used to find the current in the circuits on the opposite page?**

**Q15.** **What is the current through the generator in the first circuit on the opposite page?**

**Q16.** **What rule helps you check the calculation of the equivalent resistance of parallel resistors?**

**Q17.** **Two equal resistors in parallel have a total resistance of one half the value of one resistor. If you have a 1-ohm resistor and a 2-ohm resistor in parallel, does this rule apply?**

**Q18.** **What is the total resistance of two 7-ohm resistors in parallel?**

**Q19.** **An electric heater, an electric iron, and a lamp are fed from the same outlet (connected in parallel). Find the equivalent resistance if you know that the individual resistances are 4 ohms, 150 ohms, and 30 ohms respectively. Begin by drawing a schematic diagram of the circuit.**

**Your Answers Should Be:**

**A14. Yes,** Ohm's law can be used.

**A15.** $I = \frac{E}{R} = \frac{10}{1.2} = 8.3$ **amperes**

**A16.** When you calculate the equivalent resistance of a group of parallel resistors, the result must be smaller than the smallest single parallel resistor.

**A17. No,** the rule does not apply to resistances that are not equal.

**A18. 3.5 ohms.**

**A19.** Your schematic should look like this:

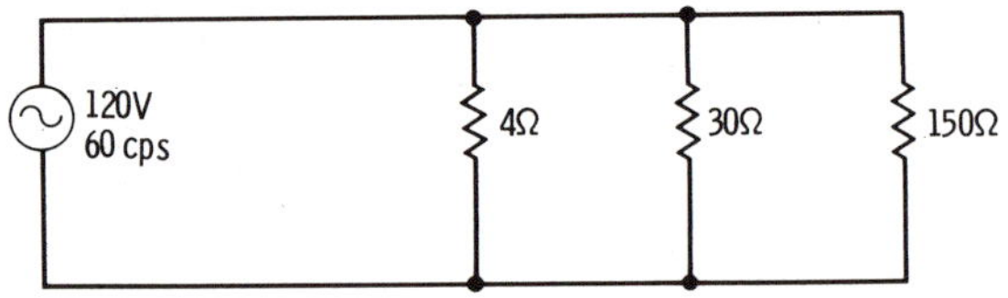


You can simplify this as follows:

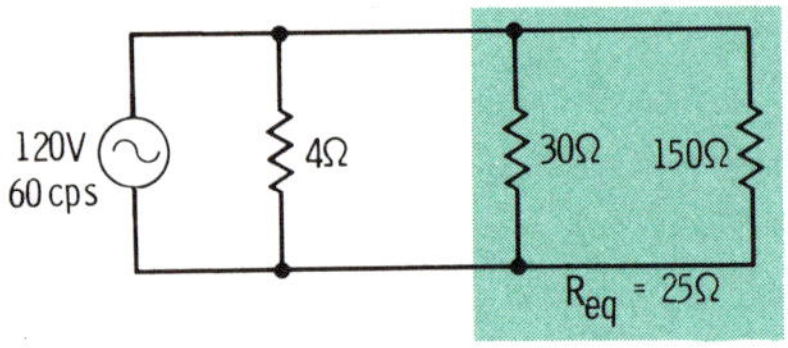


$$R_{eq} = \frac{4 \times 25}{4 + 25} = \frac{100}{29} = 3.44 \text{ ohms}$$

## AC CIRCUITS WITH RESISTANCES IN SERIES AND PARALLEL

Now that you have learned how to handle resistors in series and pairs of resistors in parallel, it is easy to solve combinations by breaking them down into simple groups of resistors in series or pairs of resistors in parallel.

Here is an example of how to break down a combination of series and parallel resistances.

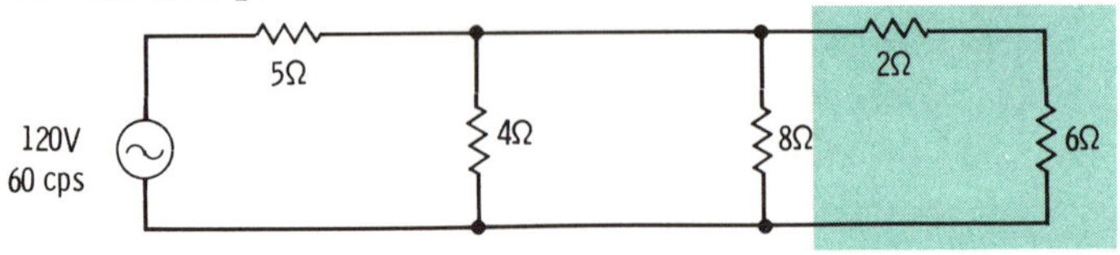

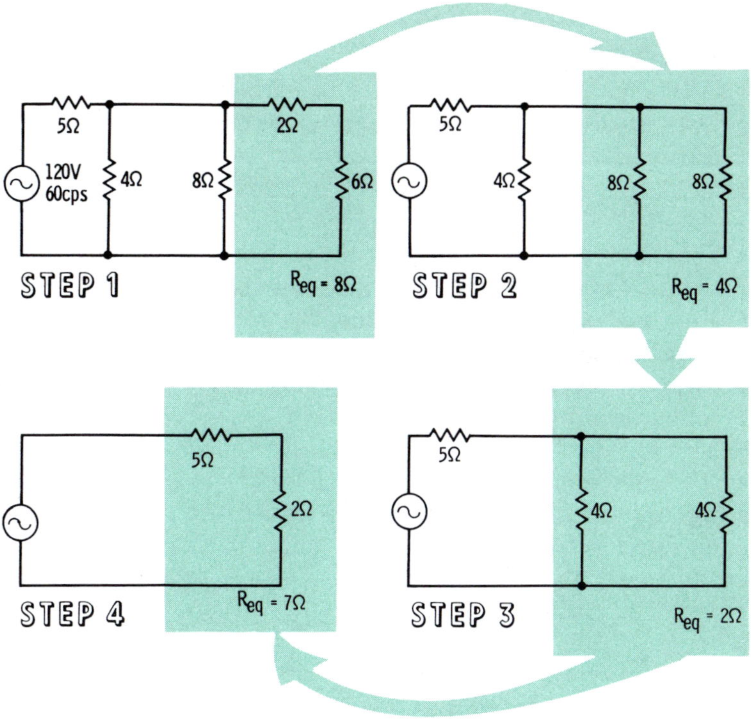


**Q20. Find the equivalent resistance of the following circuit.**

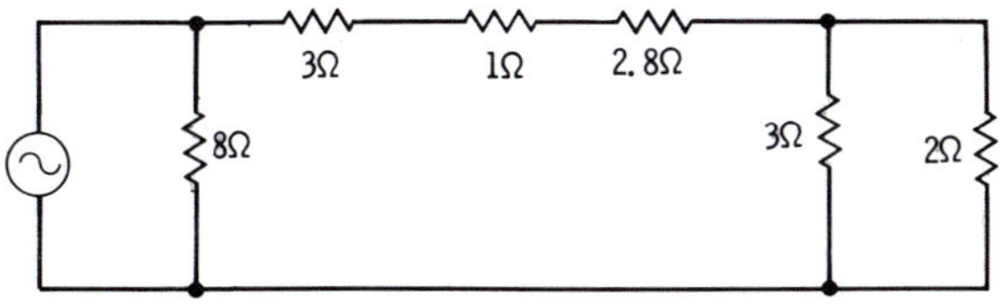


**Q21. If you calculated the equivalent resistance of the following circuit and got 6 ohms, would you be satisfied?**

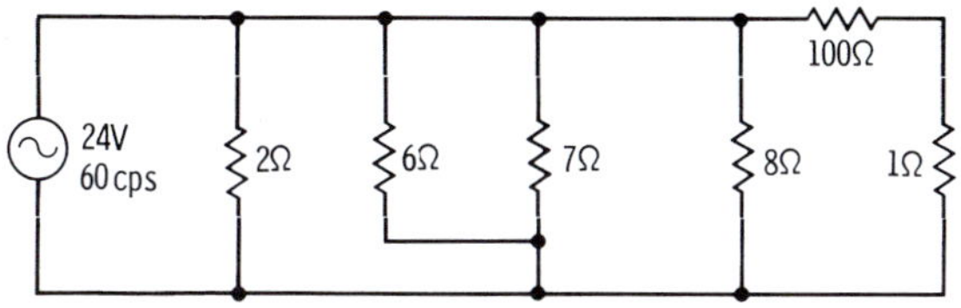

**Your Answers Should Be:**

**A20.** **4 ohms.**

**A21.** **No.** A 2-ohm resistance in parallel with the rest of the circuit means $R_{eq}$ must be less than 2 ohms.

## SKIN EFFECT WITH HIGH FREQUENCY

You have seen how an AC circuit containing only pure resistance is treated exactly the same as a DC circuit containing the same resistance values.

Usually a resistor represents the same value of resistance in both AC and DC circuits. You will find, however, that this becomes less and less true as the frequency is increased. When you deal with frequencies in the megacycle (million cycles per second) range, you will see that a resistor has a higher value. This is due to **skin effect.** At very high frequencies, electrons tend to flow only on the "skin" of a conductor.

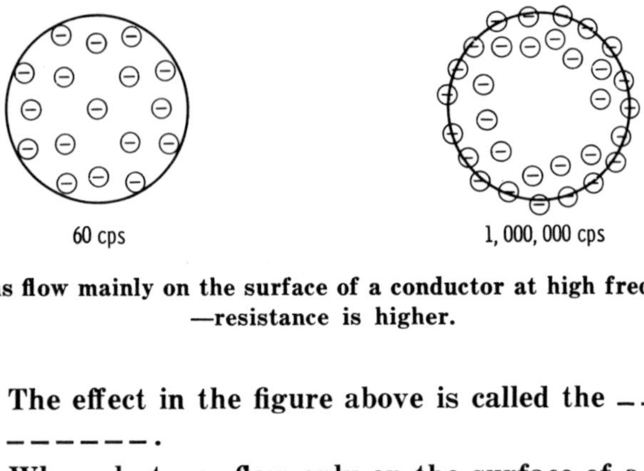


**Electrons flow mainly on the surface of a conductor at high frequencies —resistance is higher.**

**Q22.** **The effect in the figure above is called the ____ ______.**

**Q23.** **When electrons flow only on the surface of a conductor, it is as though the conductor were smaller. Resistance is (higher, lower).**

**Q24.** **If a resistor has a value of 10 ohms when measured with an ohmmeter, what would you expect its resistance to be in a circuit using 2-megacycle AC?**

**Q25.** **At high frequencies of AC current, resistances become ______ because of ____ effect.**

## WHAT YOU HAVE LEARNED

1. You have learned how to draw the schematic of a basic AC circuit.
2. Calculations with Ohm's law, or the power formulas, use rms values of voltage and current.
3. Voltage and current sine waves are in phase when they vary in the same way at the same time.
4. Voltage and current are always in phase in an AC circuit that contains only resistance.
5. Voltage and current must be in phase when they are used together in a single formula.
6. The equivalent resistance of combinations of series and parallel resistors can be found in the same way for AC circuits as for DC circuits.
7. At very high frequencies the resistance of a conductor increases because the electrons tend to flow mainly on the surface of the conductor. This is called skin effect.

**Your Answers Should Be:**

A22. This effect is called the **skin effect.**

A23. Resistance is **higher.**

A24. If a resistor has a value of 10 ohms when measured with an ohmmeter, **its resistance in a circuit using 2-megacycle AC current should be more than 10 ohms.**

A25. At high frequencies of AC current, resistances become **higher** due to the **skin** effect.

# 9

# Inductance

## *What You Will Learn*

**Inductance is one of the most important properties in electricity and electronics. Relays, transformers, coils, and many other devices all depend on inductance for their operation. When you have finished this chapter, you will know what factors influence inductance and how the inductance of a circuit affects AC voltage and current.**

### WHAT IS INDUCTANCE?

When current begins to flow in a conductor, a magnetic field builds up around it. As the magnetic field builds up, its expanding lines of force cut the conductor and generate a voltage that opposes the increasing current.

This opposing voltage, or **counter emf,** is greater when the current is changing more rapidly. In fact, the counter emf is proportional to the **rate of change** of the current, but it always opposes it. When current is decreasing, the counter emf attempts to keep the current flowing.

When a sine-wave current flows in an inductor (coil), the current is continually changing. Notice below that it is

**Current at the peaks is steady for an instant.**

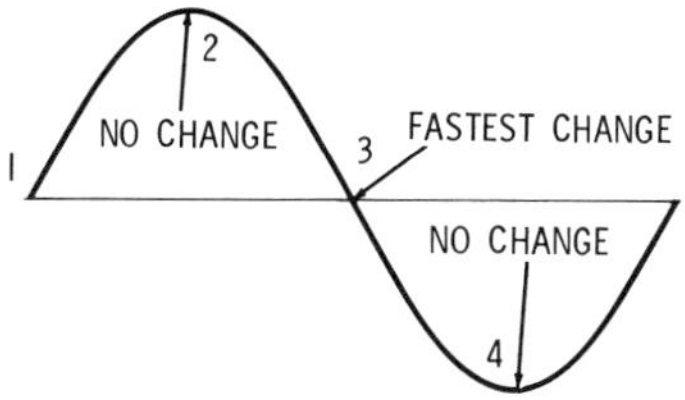

changing faster at the points where the sine wave crosses the zero line. It is not changing at all at the **instant** of each positive and negative peak.

The voltage in the inductor follows the rule just given. At point 1 in the illustration, the current is rising at its fastest rate. Therefore the counter voltage, trying to keep the current from increasing, is at its negative peak. At point 2 on the wave, the current is not changing at all; at this point the counter voltage is zero. At point 3 the current is decreasing at its maximum rate, so the counter voltage, in trying to keep the current from decreasing, reaches its positive peak. At point 4 the current is at its negative peak and is not changing at all—counter voltage is zero.

We can follow the current sine wave point by point and, at every instant, calculate its rate of change and the resulting counter emf. The resulting voltage waveform is another sine wave, but this one is 90° out of phase with the current. This is the waveform of the **counter emf.**

In order to keep the current flowing, an external voltage that is exactly equal but opposite to the counter emf must be applied. This is the applied emf, and it is 90° ahead of the current.

We say that **in an inductance, the current sine wave lags the applied voltage wave by 90°.**

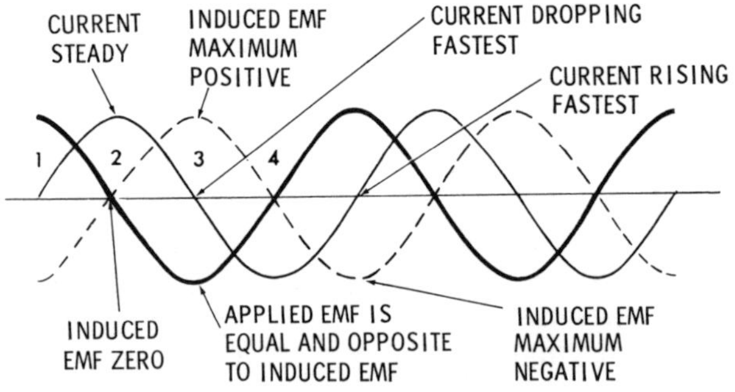


**Current lags the applied voltage by 90°.**

In a DC circuit, inductance has an effect only when the direct current first starts to flow, and then again when you try to stop it. But in AC circuits the voltage is constantly changing and inductance constantly acts to retard the change in current.

All conductors have some inductance. Straight wires have very small amounts, while coils have much more. In formulas, inductance is represented by the letter **L.** A coil or inductor is indicated on diagrams by one of the following.

**Symbols for inductance.**

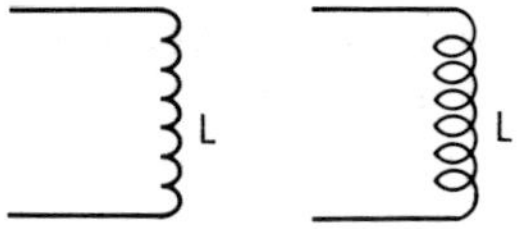


The unit of inductance is the **henry.** A coil is said to have an inductance of 1 henry if the current through it, changing at a rate of 1 ampere per second, encounters an opposition, or counter voltage, of 1 volt. This means that the **opposition to current change** shows up as a voltage opposing the applied voltage.

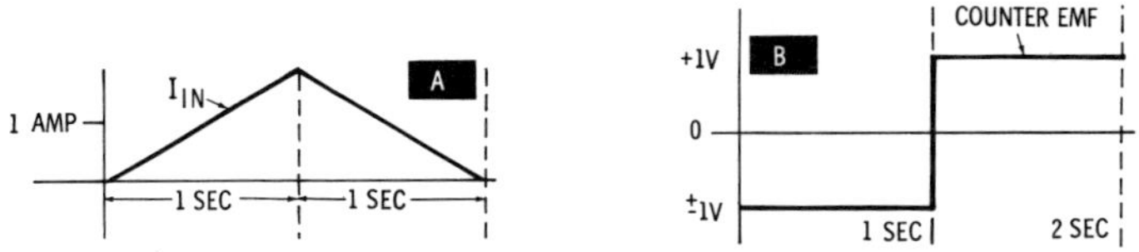


**Changing current produces a counter emf.**

In part A above, with the current **increasing** at 1 ampere per second, a counter voltage of 1 volt appears and opposes the increase in current. In part B, as the current is **decreasing,** the inductive counter voltage of 1 volt is in a direction that tends to keep the current flowing. One henry is a very large value of inductance. Therefore, inductances in millihenrys (mh) and in microhenrys ($\mu$h) are more often found.

**Q1. When current is trying to increase, inductance (makes it increase more quickly, slows down the increase).**

**Q2. Inductance opposes a change in ________.**

**Q3. Which kind of current will be most affected by inductance, AC or DC?**

**Q4. The usual symbol for inductance in formulas is the letter ____.**

**Q5. What units are used to measure the inductance of a coil?**

**Your Answers Should Be:**

**A1.** When current is trying to increase, inductance **slows down the increase.**

**A2.** Inductance opposes a change in **current.**

**A3.** **AC** is most affected by inductance.

**A4.** The usual symbol for inductance in formulas is **L.**

**A5.** **Henrys, millihenrys,** and **microhenrys** are units used to measure inductance.

## HOW DOES INDUCTANCE AFFECT AC?

If a **sine-wave** voltage is applied across a resistor, the current through the resistor is also a sine wave. At every instant of the voltage wave, the current is determined by Ohm's law and equals E/R. The two sine waves, voltage and current, are exactly in step; they are said to be **in phase.**

Inductance **resists a change in current.** But the voltage of a sine wave is always changing, and therefore is always trying to change the current through an inductance. This means that inductance acts at all times in an AC circuit and **retards** the change in the current. This results in a **current wave** that is delayed after the applied voltage wave. The current wave **lags** the voltage wave by exactly 90°, or one quarter of the period of the sine wave. The two waves are **out of phase** by 90°.

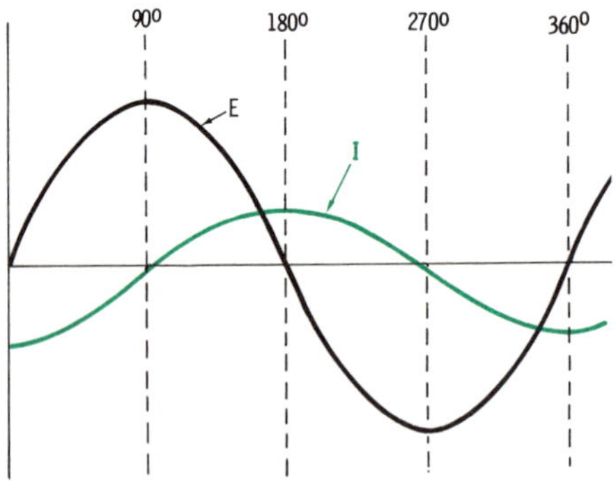


**Current lags the voltage by 90° in an inductance.**

In a circuit containing only resistance, the voltage and current are in phase, and the voltage and current vectors have the same position. In a circuit having only inductance, the current vector is 90° behind the voltage vector. The length of each vector represents its magnitude.

**Current vector lags voltage vector by 90°.**

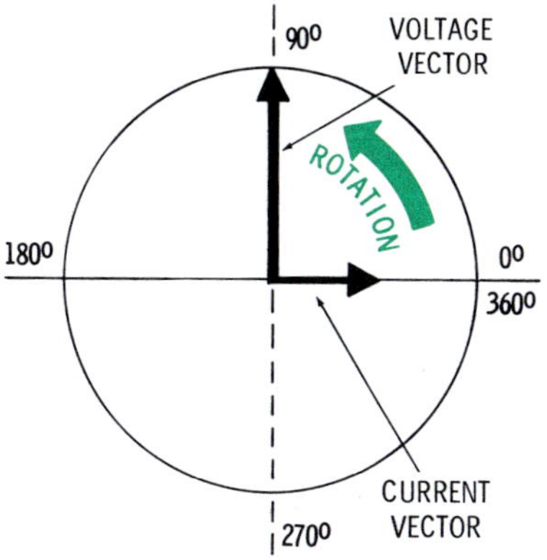


Inductance, unlike resistance, **consumes no power.** When the current in the circuit is **increasing,** inductance **takes energy out of the circuit.** It converts this energy into a magnetic field. When the current in the circuit is **decreasing,** however, this magnetic field collapses, and **all the energy returns to the circuit.** Energy is borrowed, but none is used.

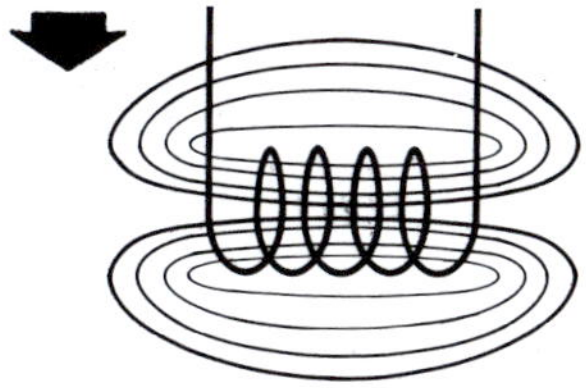

**Current increases—electricity produces magnetism.**

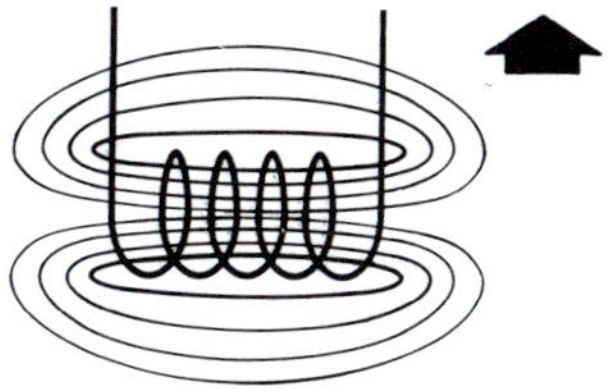

**Current decreases—magnetism produces electricity.**

**Q6. The current sine wave in an inductance (leads, lags) the voltage sine wave.**

**Q7. Current lags voltage by one-quarter cycle, or _____.**

**Q8. In an inductor, applied _ _ _ _ _ _ _ leads _ _ _ _ _ _ _ by 90°.**

**Q9. How much power is used in a pure inductive circuit?**

**Q10. An inductor stores electrical energy by producing a _ _ _ _ _ _ _ _ _ _ _ _ _ _.**

**Your Answers Should Be:**

**A6.** The current sine wave in an inductor **lags** the voltage sine wave.

**A7.** Current lags voltage by one-quarter cycle, or **90°.**

**A8.** In an inductor, **voltage** leads **current** by 90°.

**A9.** **No power** is consumed in a circuit containing only inductance.

**A10.** An inductor stores electrical energy by producing a **magnetic field.**

## FACTORS INFLUENCING INDUCTANCE VALUE

You have learned that inductance is a property of a circuit or of a component, and that a coil is the component with the most inductance.

Several factors determine the amount of inductance in a coil. One of the most important factors is the number of turns in the coil. The inductance of a coil is proportional to the **square of the number of its turns.** This means that if a certain coil has twice as many turns as another, it will have four times as much inductance; three times as many turns, it will have nine times as much inductance, etc.

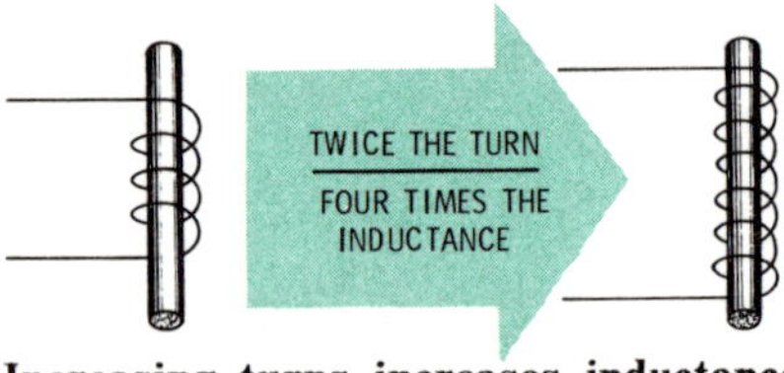


**Increasing turns increases inductance.**

The diameter also affects the inductance of a coil. **The larger the diameter, the more inductance it will have.**

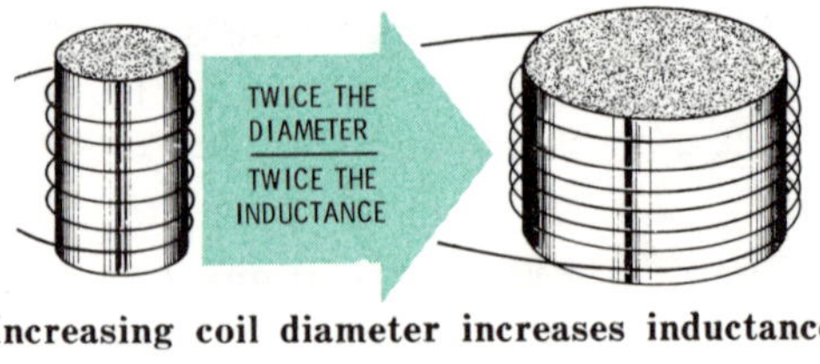


**Increasing coil diameter increases inductance.**

Placing an iron core in the center of a coil is another way to increase inductance. **A coil wound on an iron rod has much more inductance** than an air-core coil. This is because an iron core can sustain a much greater magnetic field than air and, as you have learned, the inductance of a coil is related to the amount of magnetism it can produce.

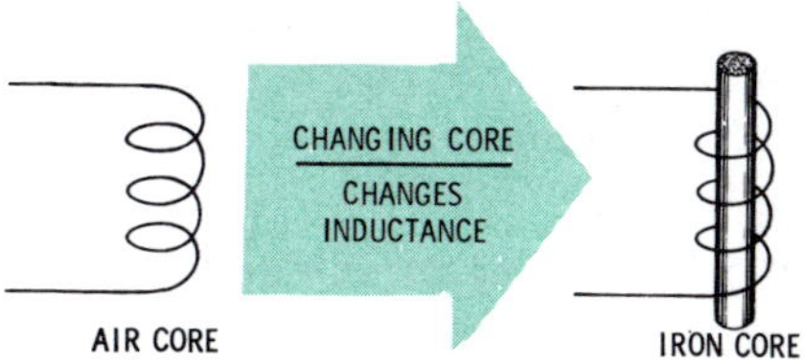


**Changing core material changes inductance.**

There are formulas for calculating the inductance of various types of coils. There are also tables for simple, one-layer coils. Using these formulas, you can design a coil to have any desired value of inductance, or calculate the value of an unknown inductance.

**Q11. The diameter of a coil and the kind of core it has are two factors that influence the amount of inductance a coil has. Name another important factor.**

**Q12. A coil with a large number of turns generally has (more, less) inductance than a coil with fewer turns.**

**Q13. Which has a greater inductance, an iron-core coil or an air-core coil? Why?**

**Q14. Which of the coils below do you think will have the greater inductance?**

**Q15. Name three factors that influence the inductance of a coil.**

**Your Answers Should Be:**

**A11.** Another important factor influencing the inductance of a coil is **the number of turns it has.**

**A12.** A coil having a large number of turns generally has **more** inductance than a coil with fewer turns.

**A13.** An **iron-core coil** has a greater inductance **because the iron-core sustains a greater magnetic field and the coil can store more electrical energy in the magnetic field.**

**A14.** **Coil B** probably has more inductance than coil A because it has an iron core.

**A15.** Three factors that influence the inductance of a coil are: **diameter, number of turns,** and **type of core.**

## INDUCTANCE AND INDUCTION

Inductance is closely related to **induction.** Inductance is a circuit property. Induction, on the other hand, is the interaction between electric current and a magnetic field. Whenever a current flows in a conductor, its sets up a magnetic field around the conductor. This is how solenoids and electromagnets work.

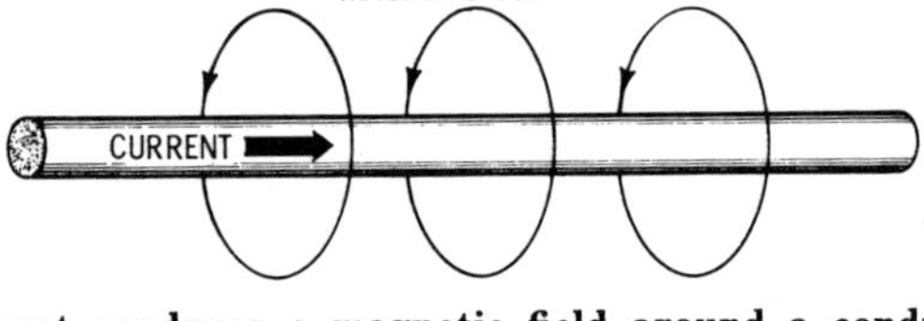


**Current produces a magnetic field around a conductor.**

A good way to remember the direction of the induced magnetic field is the **left-hand rule.** With your left hand grasping the wire and your thumb pointing in the direction of the current, the curved fingers of your hand indicate the direction of the field. The direction of the magnetic field is always the direction toward the north-seeking pole of the magnet.

A coil with a direct current flowing through it in a particular direction acts as a magnet with a fixed polarity, just as if it were a bar magnet. When the current is AC instead of DC, the polarity of the magnetic field alternates in the same manner as the current.

Conversely, if an electric conductor is moved through a magnetic field, an electric current is **induced** in the conductor. This is how generators work.

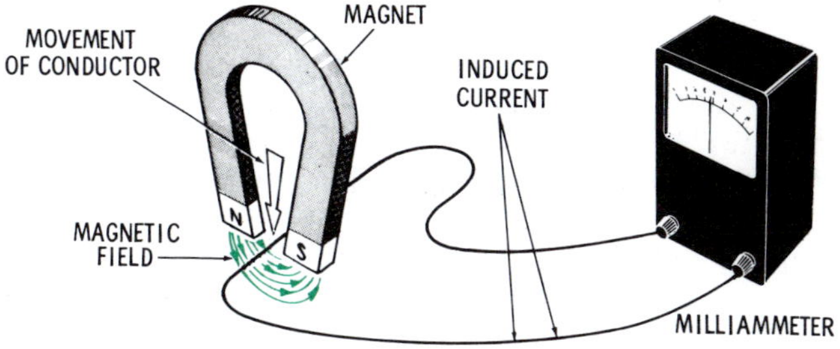


**A magnetic field induces current in a moving conductor.**

If a coil is connected to an ammeter and a bar magnet is moved through the coil, the ammeter will show that an electric current flows. This current is **induced** by the magnetic field only. If you move the bar magnet back and forth through the coil continuously, the induced current will be AC. Use the lowest amperage range on the multimeter for this experiment.

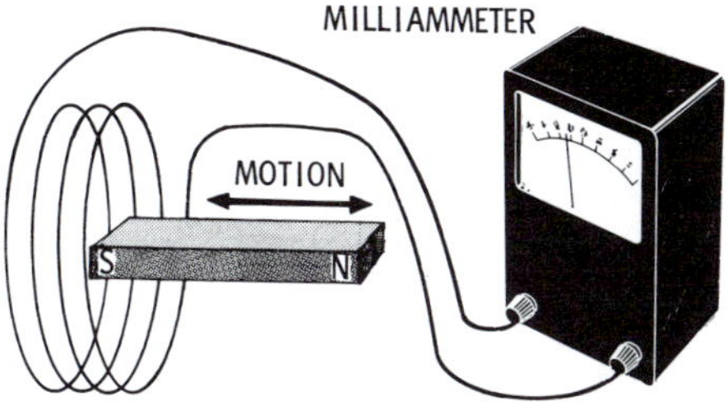


**A magnet moving in a coil produces an AC current.**

**Q16. An electric current flow produces a _________ _____.**

**Q17. Moving a conductor in a magnetic field ________ a current in the conductor.**

**Q18. Inductance is opposition to a(an) ______ __ _________.**

**Your Answers Should Be:**

**A16.** An electric current flow produces a **magnetic field.**

**A17.** Moving a conductor in a magnetic field **induces** a current in the conductor.

**A18.** Inductance is opposition to a **change in current.**

## INDUCTIVE REACTANCE

When current in an inductance is changing, the inductance opposes the change by generating a counter emf. In a DC circuit, this effect is present only at the time a switch is closed or opened, and it dies away in a few moments.

In AC circuits, however, current is constantly changing, so inductance is constantly acting to oppose it. The faster the current changes, the more opposition there will be. Obviously, the higher the frequency of a sine wave, the faster the current will change. Inductance, therefore, tends to offer more opposition at high frequencies than at low ones.

In an AC circuit this reaction to a changing current is present in addition to ordinary resistance. This opposition to the flow of an AC current through an inductance is called **inductive reactance.** The greater the inductance of the circuit, the greater is the inductive reactance. Inductive reactance does **not** oppose the flow of DC current (zero frequency).

The more rapidly the current is changing, the greater the opposition, or inductive reactance, will be. Since the rate of change in the current depends on the frequency of the sine wave, the higher the frequency, the greater the opposition to current flow will be. The symbol for inductive reactance is $\mathbf{X_L}$.

The formula for inductive reactance is:

$$X_L = 2\pi fL \ (\pi = 3.14)$$

Inductive reactance ($X_L$) is measured in **ohms,** as resistance is (but don't confuse the two). In the above formula, L is the inductance in henrys, and f is the frequency in cycles per second.

From the formula, it is apparent that $X_L$ increases with frequency. When the frequency is doubled, $X_L$ doubles. Why is this? Because when the frequency is doubled, the current is reversing twice as fast, and the opposition to this change (caused by the inductance) also doubles. Notice, too, that if the frequency in the formula is equal to zero, inductive reactance disappears completely. A DC current has zero frequency and is not affected by inductance.

Inductance is very useful because every inductive circuit is **frequency-sensitive.** This principle is used in filters, antennas, and many other applications. It means that an inductive circuit passes direct current and low-frequency alternating current, but it impedes the higher frequencies.

**Q19. The opposition of a coil to the flow of AC current is called ---------- ----------.**

**Q20. In what kind of units is $X_L$ measured?**

**Q21. Inductive reactance depends on the value of inductance and ----------.**

**Q22. How is the inductance of a circuit affected by the input signal?**

**Q23. How is the inductive reactance of a circuit affected by the input signal?**

**Q24. What is the inductive reactance of the coil in the following circuit?**

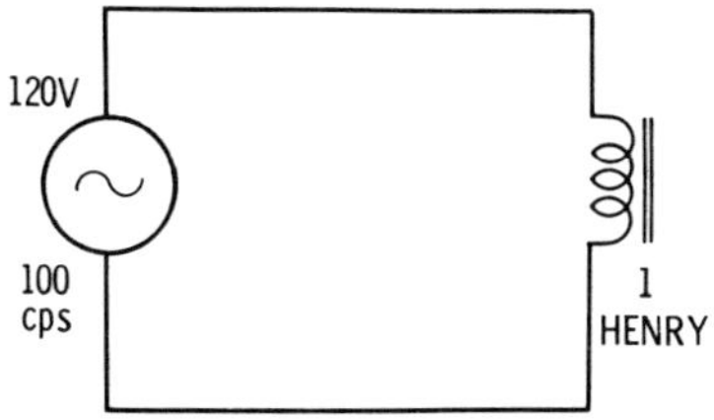


**Q25. How much current is flowing in the above circuit?**

**Q26. If the frequency of the AC source in the above circuit is doubled, the inductive reactance will (decrease, increase, remain the same).**

**Q27. If an ohmmeter is used to measure a coil, the reading will indicate (DC resistance, inductive reactance).**

**Your Answers Should Be:**

**A19.** The opposition of a coil to the flow of AC current is called **inductive reactance.**

**A20.** $X_L$ is measured in **ohms.**

**A21.** Inductive reactance depends on the value of inductance and **frequency.**

**A22.** Inductance is **not affected** by the input signal.

**A23.** Inductive reactance **increases as the frequency of the input signal increases.**

**A24.**
$$X_L = 2\pi fL = 2 \times 3.14 \times 100 \times 1 = \mathbf{628\ ohms}$$

**A25.**
$$I = \frac{E}{X_L} = \frac{120}{628} = \mathbf{0.191\ ampere}$$

**A26.** If the frequency of the AC source is doubled, the inductive reactance will **increase.**

**A27.** If an ohmmeter is used to measure a coil, the reading will indicate **DC resistance.**

## APPLICATION OF INDUCTANCE

Because inductive reactance depends on frequency, inductance is often used in **filters.** Filters are special circuits which have the property of allowing certain frequencies to pass while blocking others. There are, for example, **low-pass filters** which pass low frequencies and block high ones, and **band-pass filters** which pass only a certain band of frequencies.

Here are two simple filters that depend only on inductance. The first one, which has an inductance in series, blocks high frequencies. It is a **low-pass** filter.

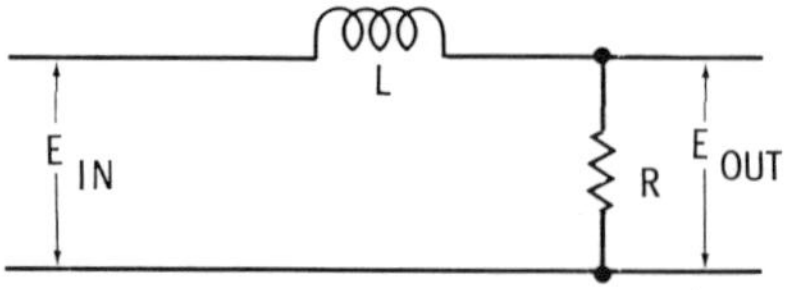


**A low-pass filter.**

The second circuit has an inductance in parallel, or across it. This inductance will **bypass** the low frequencies. As the frequency increases, voltage across the inductor increases. Therefore, there is more output voltage at high frequencies, and the circuit acts as a **high-pass** filter.

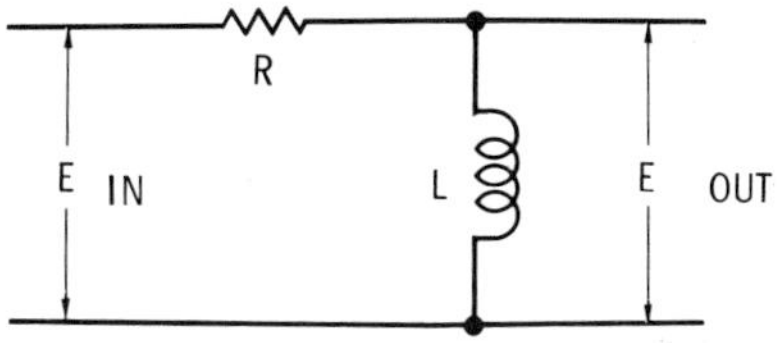


**A high-pass filter.**

If you have a signal generator, you can build a simple filter circuit like the one shown below and see how it reacts to different frequencies.

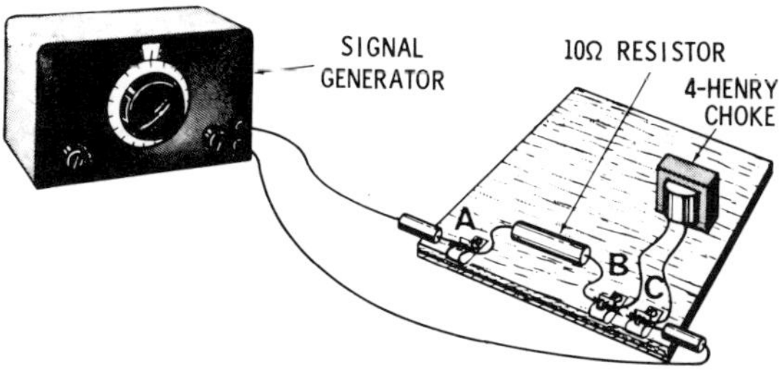


**Filters use inductance.**

Voltages can be measured between points A and C, A and B, and B and C. Pick about 12 equally spaced frequencies. Using each as an input, measure and record both the input and output voltages at each frequency.

**Q28. A filter that allows only high frequencies to pass is called a ---- ---- ------.**

**Q29. A ---- ---- ------ is designed to allow only a certain band of frequencies to pass.**

**Q30. Draw a schematic of a simple low-pass filter.**

**Q31. Between which points would you measure the input voltage to the filter circuit in the second illustration above?**

**Q32. Between which points would you measure the low-pass output of the filter?**

**Q33. Between which points would you measure the high-pass output of the filter?**

**Your Answers Should Be:**

**A28.** A filter that allows only high frequencies to pass is called a **high-pass filter.**

**A29.** A **band-pass filter** is designed to allow only a certain band of frequencies to pass.

**A30.** Your schematic should look like this:

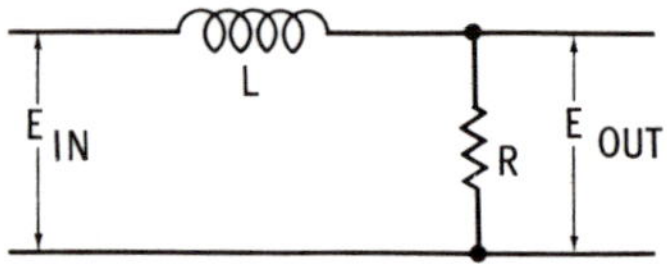


**A31.** Measure input voltage **between points A and C.**

**A32.** Measure the low-pass output **between A and B.**

**A33.** Measure the high-pass output **between B and C.**

## TRANSFORMERS

You are now aware that a moving magnetic field generates an electric current in a conductor, and also that current flowing in a conductor produces a magnetic field. These two effects can be combined in a circuit such as this.

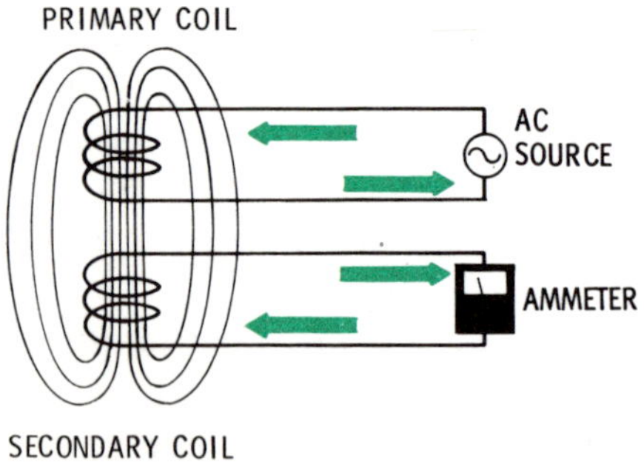


**The transformer principle.**

One coil has a current flowing in it. It is an AC current that sets up an alternating magnetic field in and near the coil. If another coil is placed next to it, there will be a second alternating current induced in the second coil. The first coil is the **primary,** the second coil is the **secondary,** and the combination of the two is a **transformer.** Most commercial transformers appear as shown on the next page.

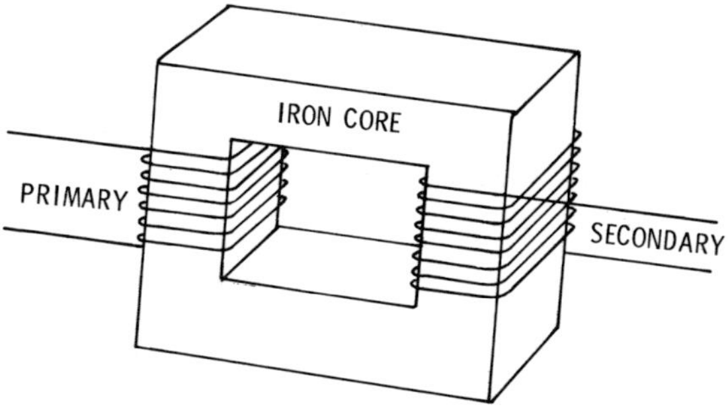


**Most transformers have an iron core.**

An iron core is used to increase the magnetic flux and to channel it to the secondary coil. The primary coil sets up a magnetic field in the core, and the secondary coil converts the field back to electric current. Power is actually transferred from the primary to the secondary. A lamp or other load placed in the secondary circuit will operate.

One of the main advantages of using transformers is that they can **change voltage.** They do this because the voltage induced in the secondary depends on the number of turns in the secondary as compared to the number of turns in the primary coil. If the turns in the secondary are doubled, the induced voltage will also be doubled (but no more power, because the current will be halved). The **voltage ratio** of the secondary to primary is the same as the **turns ratio.** So, if the primary of a transformer has 1,000 turns and the secondary has 100 turns, it is a **step-down transformer** because it steps down the primary voltage by 10 (1,000/100). If the connections are reversed, it becomes a **step-up transformer** with the same ratio.

**Q34. In the primary coil of a transformer, electric current produces a _________ ______.**

**Q35. In the secondary coil of a transformer, a moving magnetic field produces a(an) _________ ________.**

**Q36. How does a transformer work?**

**Q37. Will a transformer work with DC electricity?**

**Q38. If a transformer has 100 turns in one coil and 200 turns in the other, how could you get a 240-volt output using ordinary household current?**

**Your Answers Should Be:**

**A34.** In the primary coil of a transformer, electric current produces a **magnetic field.**

**A35.** In the secondary coil of a transformer, a moving magnetic field produces an **electric current.**

**A36.** An AC current produces a changing magnetic field in the primary coil of a transformer. This changing field produces an AC electric current in the secondary coil of the transformer.

**A37.** **No.** A transformer will work only with AC.

**A38.** **Apply the 120V household current to the 100-turn coil** and you will get 240V from the 200-turn coil.

## PULSE RESPONSE

When pulses are applied to a circuit containing an inductor, the inductor opposes a change of current in its usual way. The effect of this is to distort the waveform of the current through the inductor by rounding off the corners on the leading and trailing edges.

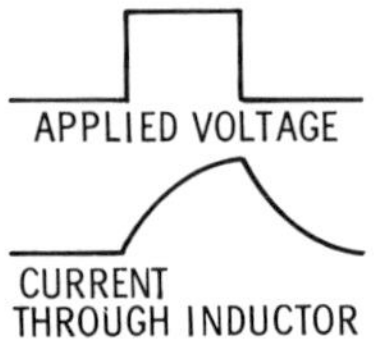


**An inductor rounds off a current waveform.**

The voltage across the inductor is affected in just the opposite way. An increase in current tends to cause a sharp increase of voltage in the **positive** direction. A decrease in current causes an increase of voltage in the **negative** direction.

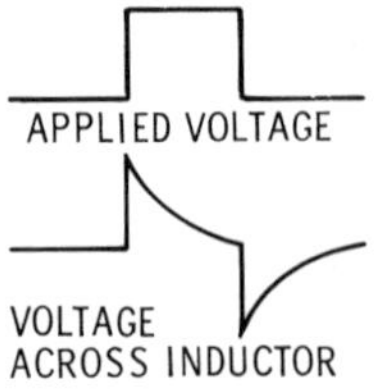


**An inductor exaggerates a voltage waveform.**

Inductive circuits are used in practically all electronic equipment. The frequency-sensitive characteristics of an inductor find practical uses in a TV receiver, for example. High video frequencies are not amplified as much as low video frequencies as they pass through the various stages of the set. However, the quality of the picture displayed on the screen depends to a large extent on an equal amplification of both the high and low frequencies. **Peaking** coils are therefore used in the better-quality receivers to equalize the amplification of high and low frequencies. The high frequencies cause a larger voltage drop across the peaking coils than the low frequencies. This tends to compensate for the unbalance in amplification, resulting in a better picture.

Inductors are also used in some hi-fi speaker systems to improve sound reproduction quality. The low audio frequencies are separated from the high audio frequencies by an inductive filter circuit. The lows are then fed to the larger speakers (woofers) and the highs to the smaller speakers (tweeters). Thus, each speaker reproduces only those frequencies which it can handle without distortion. This results in a more faithful and distortion-free reproduction of the amplified sound.

**Q39. Inductance opposes a change in current. It tends to round off the corners on both edges of the ______ waveform of a pulse.**

**Q40. The effect of inductance on the voltage waveform of a pulse is the opposite of its effect on the current waveform; it will turn a corner into a sharp spike in the ______ waveform.**

**Q41. In the figure at the bottom of the opposite page, what happens to the inductor voltage when the current increases?**

**Q42. What happens to the inductor voltage when the current decreases?**

**Q43. Why must a TV receiver amplify the high video frequencies as much as it amplifies the low video frequencies?**

**Q44. What is the name of the components that are used to equalize the gain for high and low video frequencies in many TV receivers?**

**Your Answers Should Be:**

**A39.** Inductance opposes a change in current. It tends to round off the corners on both edges of the **current** waveform of a pulse.

**A40.** The effect of inductance on the voltage waveform of a pulse is the opposite of its effect on the current waveform; it will turn a corner into a sharp spike in the **voltage** waveform.

**A41.** The voltage increases in the **positive** direction.

**A42.** The voltage increases in the **negative** direction.

**A43.** The gain must be uniform **to provide a good-quality picture.**

**A44.** **Peaking coils** are used to equalize the video gain.

## WHAT YOU HAVE LEARNED

1. Inductance opposes a change in current and makes the waveform of an AC current lag the voltage by 90°.
2. Inductance is measured in henrys, millihenrys, or microhenrys.
3. Inductance uses no power; it only "borrows" energy for a short time.
4. Three factors that influence the inductance of a coil are number of turns, diameter of the coil, and type of core.
5. Induction is the creation of a magnetic field by an electric current or of an electric current by a magnetic field.
6. Transformers work by induction.
7. Inductive reactance acts similarly to resistance in an AC circuit and is measured in ohms.
8. The formula for inductive reactance is:

$$X_L = 2\pi fL$$

9. You have learned how to use inductance to construct a simple filter circuit.
10. You have learned how inductance affects the waveform of pulses.

# 10

# RL Circuits

## *What You Will Learn*

**You are now going to learn how to find the equivalent reactance of inductors in series and in parallel. You will become acquainted with some of the ways to measure the relative importance of R and L in a component or circuit and how to find the time constant of a circuit. You will be able to add reactance and resistance to find impedance. You will discover how the combination of resistance and inductance affects the phase relations of voltage and current. You will also learn how to calculate current and power in RL circuits.**

### INDUCTIVE CIRCUITS

A good way to understand circuits containing inductance is to work through a simple problem. For example, the following illustration shows a schematic of a basic inductive-reactance AC circuit.

**A basic inductive-reactance circuit.**

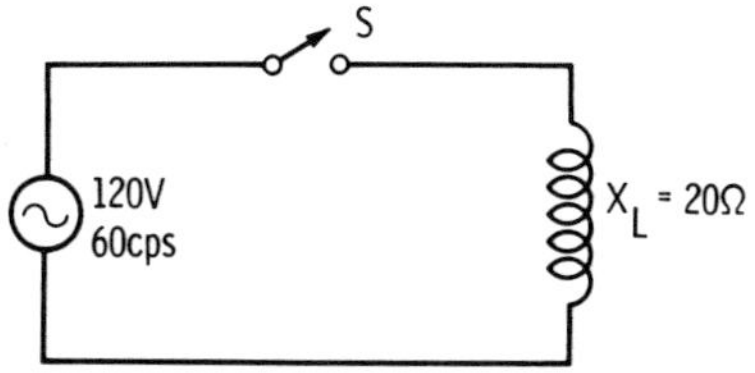


Assume that the resistance in the leads and the coil is negligible. The problem is to find the rms value of the current in this circuit.

Ohm's law still applies.

$$I = \frac{E}{X_L} = \frac{120}{20} = 6 \text{ amperes}$$

You have learned that power in a circuit is:

$$P = EI$$

where,

E is the voltage,
I is the current.

With a sine wave the power at any time during the cycle is the product of the voltage and the current at that moment. In a resistive circuit, power has a pulsating waveshape.

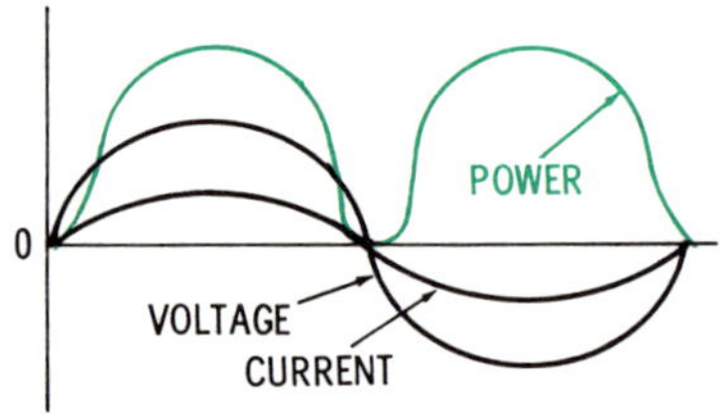


**Power in a resistive circuit.**

You have already learned that no power is dissipated in a circuit that contains only inductance. A look at the waveforms below will help you understand this.

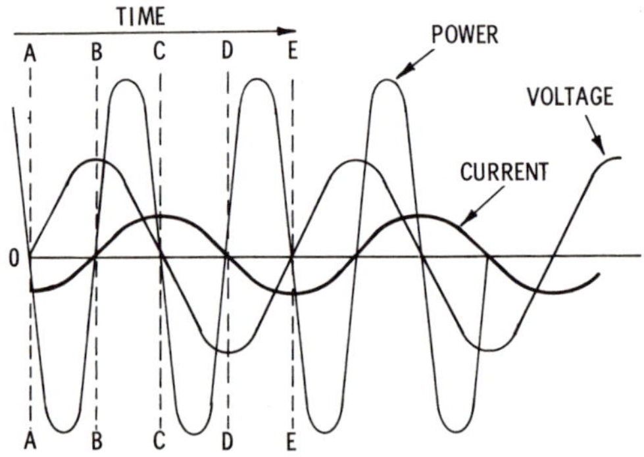


**Power in an inductive circuit.**

Between points B and C, both current and voltage are positive. If you multiply their values, it appears that power is being dissipated exactly as in a resistive circuit. Between points D and E, both current and voltage are negative, and again you have exactly the same situation as in a resistive circuit—power appears to be dissipated. But between points A and B and points C and D there is a situation that never exists in a resistive circuit.

As you can see, there are pulses of negative power as well as positive power. The positive-power pulses represent the time when the circuit is utilizing power to produce a magnetic field. The negative-power pulses represent the time when the circuit is absorbing power from the magnetic field. The negative pulses and the positive pulses are equal and cancel each other, so the total power dissipated is zero.

An important rule that you must remember is:

**When you multiply positive values by positive values, or negative values by negative values, the results are positive values.**

**When you multiply positive values by negative values, the results are negative values.**

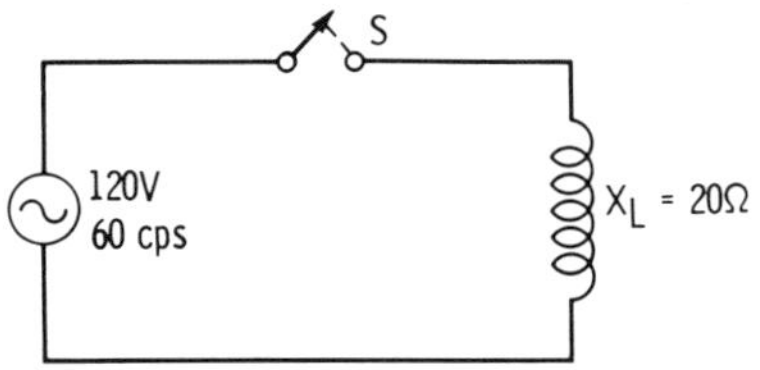


Inductive reactance in a circuit changes with frequency, but inductance stays the same. To find how the above circuit behaves at other frequencies, you must determine the inductance. Use the formula:

$$L = \frac{X_L}{2\pi f} \text{ (this is a form of } X_L = 2\pi fL)$$

**Q1. Is voltage positive or negative between points A and B in the figure at the bottom of the opposite page?**

**Q2. Is current positive or negative between A and B?**

**Q3. Is power positive or negative between A and B?**

**Q4. What is the amount of inductance in the circuit shown above?**

**Q5. How much current will flow in this circuit if the applied voltage is 100V at a frequency of 100 cps?**

**Your Answers Should Be:**

**A1.** Voltage is **positive.**

**A2.** Current is **negative.**

**A3.** Power is **negative.**

**A4. About 0.053 henry.**

**A5.** $X_L$ is 33.3 ohms.

$$I = \frac{E}{X_L} = \frac{100}{33.3} = \textbf{3 amperes}$$

## Inductors in Series

The simplest form of a series inductive circuit is one with two inductors in series.

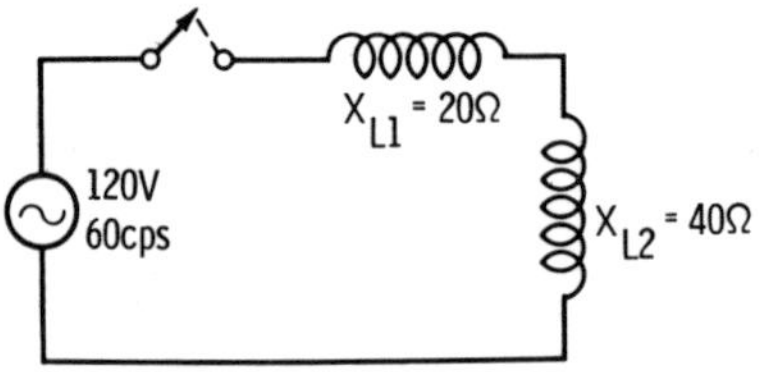


**A series inductive circuit.**

To find the current in the circuit, add $X_{L1}$ and $X_{L2}$, and then use Ohm's law with the equivalent $X_L$.

If the value of each inductance (L) is known, add the individual inductances to find the total inductance and then calculate $X_{Leq}$ for the circuit.

## Parallel Inductive Circuits

A circuit containing pure inductances in parallel can be treated much like a parallel resistance circuit.

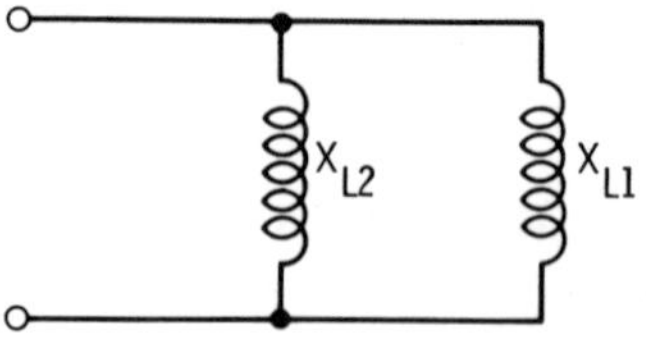


**Inductances in parallel.**

This simple parallel circuit can be solved by the formula:

$$X_{Leq} = \frac{X_{L1} \times X_{L2}}{X_{L1} + X_{L2}}$$

As with resistance, the equivalent inductive reactance of two inductors in parallel is always smaller than that of either single inductor. Combined series and parallel circuits, or large groups of inductors in parallel, can be simplified in steps by the same method you have used with resistances.

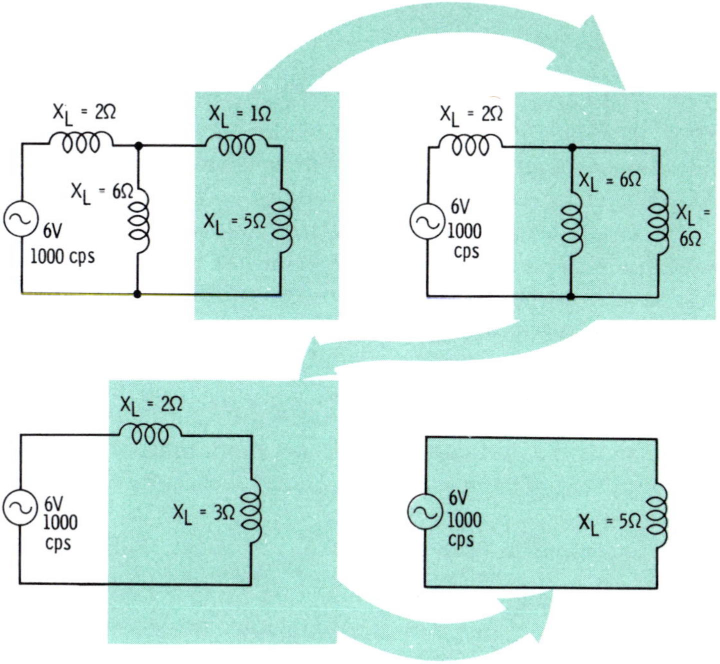


**A circuit containing series and parallel inductances can be simplified.**

**Q6. What is the equivalent $X_L$ in the following circuit?**

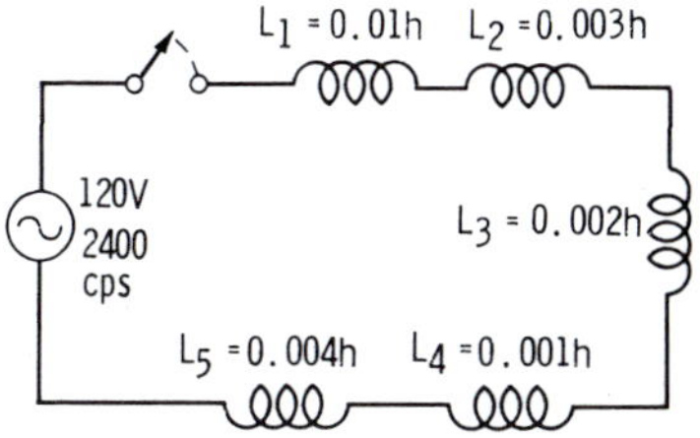


**Q7. How much current will flow in this circuit?**

**Q8. How much power will be dissipated in this circuit?**

**Your Answers Should Be:**

**A6.** $X_L = 2\pi fL = 2 \times 3.14 \times 2{,}400 \times .02 =$ **301 ohms**

**A7.** $I = \frac{E}{X_L} = \frac{120}{301} =$ **.399 ampere**

**A8. No power** will be dissipated.

## Q FACTOR

An RL circuit is one that contains both resistance and inductance. You are more likely to encounter circuits of this sort than pure inductive circuits or pure resistive circuits. In fact, even the connecting conductors in a circuit have some inductance, and every coil has some resistance.

It is sometimes important to know how "good" or how "pure" the inductance of a coil is. This quality is usually measured with a factor called **Q.** This is simply a ratio, $Q = \frac{X_L}{R}$. With a large Q, the power loss in the coil will be small, and the inductance will be more efficient. Notice, however, that Q varies with frequency, because the inductive reactance varies.

For example, a coil with an $X_L$ value of 5,000 ohms at 10,000 cps and a DC resistance of 50 ohms has a Q at that frequency of $\frac{5{,}000}{50} = 100$. If this same coil is used at 5,000 cps, its inductive reactance will only be 2,500 ohms, and its Q will equal $\frac{2{,}500}{50} = 50$.

## TIME CONSTANT

A different measure of the relative amounts of resistance and inductance is very important when dealing with pulse circuits. The shape of a pulse is changed by inductance.

When a pulse is passed through an RL series circuit, the roundness of the current rise and the time it takes for the current to rise to its final value depend on the amounts of inductance and resistance in the circuit. As you would expect, the greater the value of inductance, the slower the current will build up. At the same time, the resistance in the circuit has the opposite effect—the smaller the resistance, the longer the current takes to reach the steady-state condition. (The reason is that when the resistance is smaller, the final current will be larger, and it will therefore take longer to reach this final value.)

**Increasing the time constant increases the amount of distortion.**

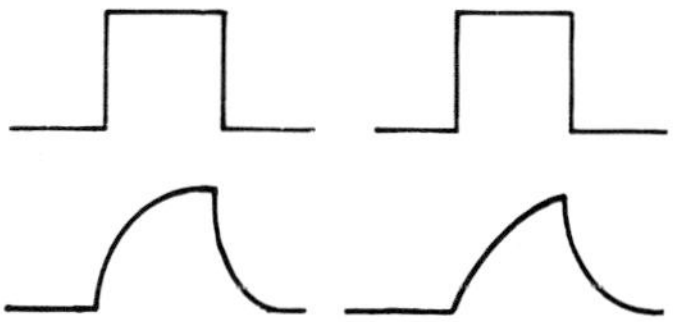

When dealing with filters and pulse circuits, circuits containing L and R in series are often described by their **time constant.** This is a measure of how quickly the current in the circuit reaches its final peak value. The time constant equals L/R and is expressed in seconds. If a circuit has an L/R time constant of one-half second, the current will reach 63% of its maximum (peak) value in one-half second when a voltage is applied to the circuit.

**L/R determines how long it takes a pulse to reach 63% of its peak.**

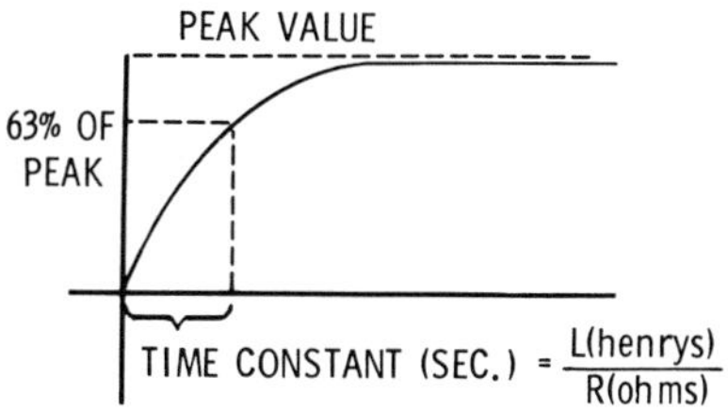


**Q9. A "pure" inductance (which has no resistance) would have a (high, low) Q.**

**Q10. If you purchased a coil with a low Q, you would expect it to have a relatively (high, low) DC resistance.**

**Q11. If a coil has a high Q at 1,000 cps, would you expect it to have a higher or lower Q at 5,000 cps?**

**Q12. If the inductance of a circuit is 3 henrys and the resistance of the circuit is 5 ohms, how long will it take for a pulse to reach 63% of its maximum value?**

**Q13. The time constant of a circuit indicates the time it takes a pulse to reach ___% of its maximum value in that circuit.**

**Your Answers Should Be:**

**A9.** A "pure" inductance would have a **high** Q.

**A10.** A coil with low Q would have a relatively **high** DC resistance.

**A11.** At an increased frequency, you would expect Q to **increase.**

**A12.** It would take a pulse $\frac{L}{R} = \frac{3}{5} =$ **0.6 second** to reach 63% of its maximum value.

**A13.** The time constant of a circuit indicates the time it takes a pulse to reach **63%** of its maximum value in that circuit.

## PHASE

It has been previously explained that AC voltage and current are always in phase in a purely resistive circuit and that AC current through an inductance always lags the applied voltage by 90°. When resistance and inductance are combined in a single circuit, the amount of phase difference between the current and voltage depends on which (resistance or inductance) has the greater value; that is, it depends on the Q of the circuit.

If the applied voltage is a sine wave, the current through an RL circuit will also be a sine wave. Therefore, you can

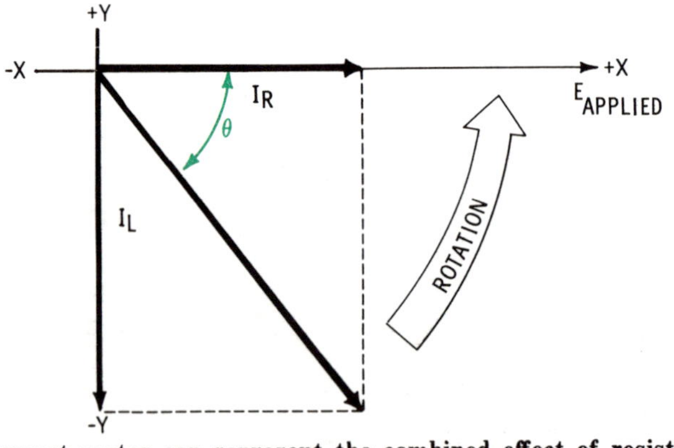


**One current vector can represent the combined effect of resistive and inductive currents.**

think of it as being generated by a rotating current vector. However, this vector is a combination (or resultant) of the resistive and inductive current vectors.

As you see, these two vectors form the two sides of a rectangle, and the overall (resultant) current vector is the diagonal of the rectangle. The angle labeled with the symbol $\theta$ is the **phase angle,** the number of degrees by which the overall current lags voltage.

## IMPEDANCE

To find the current flowing in a purely inductive circuit, you apply Ohm's law, using inductive reactance instead of resistance ($I = E/X_L$). Inductive reactance, of course, equals $2\pi fL$ and varies with frequency and inductance.

What happens when both resistance and inductance are in series in the same circuit? Say, for example, that the resistance is 3 ohms, and the inductive reactance (for a specific frequency) is 4 ohms.

As you know by now, the current through the resistance in an AC circuit is **in phase** with the applied voltage sine wave, while the current in the inductance **lags 90° behind** the voltage. Just as the rms value of resistive current cannot be added to the rms value of the inductive current to find the overall current, the 3 ohms of resistance cannot be added to the 4 ohms of inductive reactance. Instead, the overall effect of the two must be found in the same way that the overall current vector is found. The overall effect of resistance and reactance working together is called **impedance.** The symbol for impedance is **Z.**

**Q14. If the current lags the voltage by 85° in a circuit with an input at 1,000 cps, do you think an input at 450 cps is more likely to cause the current to (a) lag by 89°, (b) lead by 45°, (c) lag by 15°?**

**Q15. Draw vector lines representing 3 ohms of resistance (R) and 4 ohms of inductive reactance ($X_L$) on a sheet of paper.**

**Q16. Complete your vector diagram to find the impedance presented by the 3 ohms of resistance and 4 ohms of inductive reactance. Measure the diagonal.**

**Your Answers Should Be:**

**A14.** Current is most likely to **(c) lag by 15°.**

**A15.** You should have drawn vector lines that look like this.

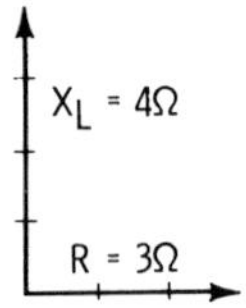


**A16.** Your vector diagram should look like this. Z is **5 ohms.**

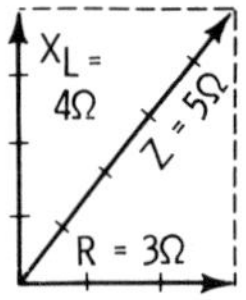


## Relationship of $X_L$, R, and Z

One simple way to find the overall effect of 3 ohms of resistance and 4 ohms of inductive reactance is to draw a line 4 units long pointing downward. This line represents the inductive reactance. Then draw a line to the right 3 units long. This line represents the resistance. The two lines form two sides of a rectangle. The diagonal of this rectangle will represent the impedance.

Notice the angle between the R and $X_L$ vectors. This angle is usually indicated by the Greek letter **theta** ($\theta$) and is referred to as the phase angle.

It is not enough to say that a circuit has an impedance of 5 ohms; you must also know the angle by which the current and voltage are out of phase. There are two ways to do this. You can express impedance in **polar** form, $Z\angle\theta$. In the example above, Z is 5. Or you can express the impedance as the sum of 3 ohms resistance plus 4 ohms inductive reactance. A short way of saying this is 3 + j4. The **j** tells you that the 4 is 90° ahead of the 3. In general

$Z = R + jX_L$. This is the **rectangular** form of impedance. Although you can find impedance by drawing vector diagrams and measuring, there are other ways of finding the value of impedance.

As long as $\theta$ remains the same, the proportion between reactance and impedance will be the same. The proportion between resistance and impedance and between reactance and resistance will also be the same.

When any two facts about a combination of $X_L$ and R are known, the other facts can be found by using a table of **trigonometric functions.** An example of such a table is shown on page 270.

Trigonometric relationships of $X_L$, R, and Z.

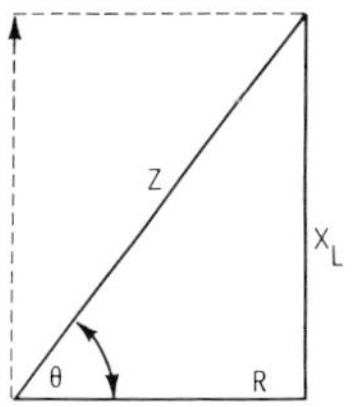


$$\tan\theta = \frac{X_L}{R} \text{ or } R = \frac{X_L}{\tan\theta}$$

$$\sin\theta = \frac{X_L}{Z} \text{ or } Z = \frac{X_L}{\sin\theta}$$

$$\cos\theta = \frac{R}{Z} \text{ or } Z = \frac{R}{\cos\theta}$$

Using the above relationships, the impedance of an inductive circuit for which $X_L$ is 7 ohms and R is 10 ohms can be found.

$$\tan\theta = \frac{X_L}{R} = \frac{7}{10} = 0.700$$

$$\theta = 35^\circ \text{ (from table on page 270) ; } \sin\theta = 0.574$$

$$Z = \frac{X_L}{\sin\theta} = \frac{7}{0.574} = 12.2 \angle 35^\circ \text{ ohms}$$

**Q17. What happens to $\theta$ if both R and $X_L$ are doubled?**

**Q18. What happens to Z if both R and $X_L$ are doubled?**

**Q19. What happens to R and $X_L$ if Z stays the same but $\theta$ is increased?**

**Q20. What is $\theta$ if R is 17.33 ohms and $X_L$ is 10 ohms?**

**Q21. What is Z for Q20?**

**Q22. If R is 12 ohms and $\theta$ is 30°, what is Z?**

**Q23. If $X_L$ is 20 ohms and $\theta$ is 30°, what is Z?**

> **Your Answers Should Be:**
>
> **A17.** If you double both R and $X_L$, $\theta$ **remains the same.**
>
> **A18.** If you double both R and $X_L$, **Z is doubled.**
>
> **A19.** If $\theta$ is increased, **$X_L$ becomes larger and R smaller.**
>
> **A20.** $\tan \theta = X_L/R = \mathbf{30^\circ}$
>
> **A21.** $Z = X_L/\sin \theta = \mathbf{20 \angle 30^\circ}$ **ohms**
>
> **A22.** $Z = R/\cos \theta = \mathbf{13.9 \angle 30^\circ}$ **ohms**
>
> **A23.** $Z = X_L/\sin \theta = \mathbf{40 \angle 30^\circ}$ **ohms**

**Current in an RL Circuit**

When inductors and resistors are connected in series in a circuit, simply add all the inductive reactances and all the resistances separately to get an equivalent circuit with one inductive reactance and one resistance.

If it is necessary to find the impedance of this circuit, begin by simplifying it.

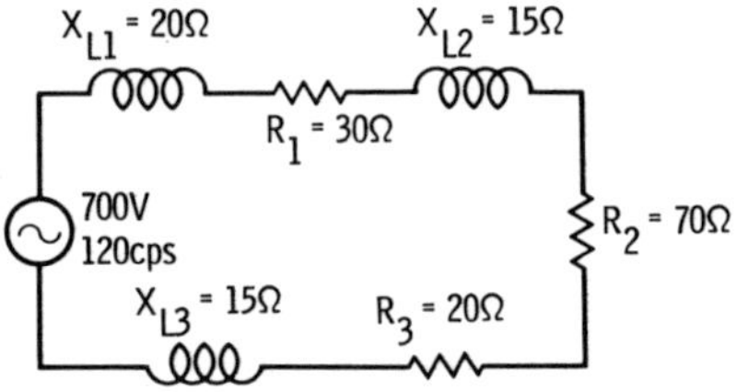


Even though $X_L$ and R values are scattered through the series circuit, they may be added directly. Thus,

$$X_{L\ TOTAL} = X_{L1} + X_{L2} + X_{L3} = 20 + 15 + 15 = 50 \text{ ohms}$$
$$R_{TOTAL} = R_1 + R_2 + R_3 = 30 + 70 + 20 = 120 \text{ ohms}$$

The circuit above is equal to this equivalent circuit.

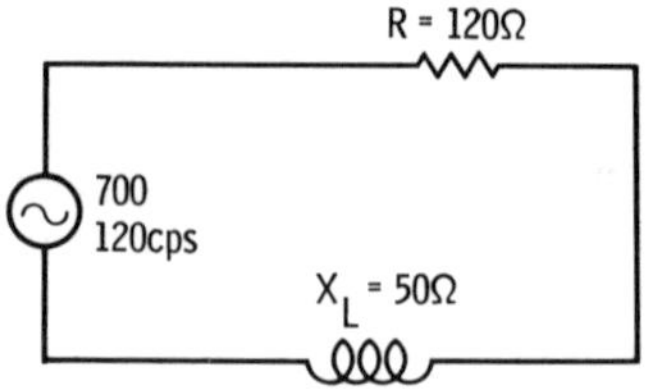

## POWER IN RL CIRCUITS

All power is dissipated in resistance. You have also learned that the formula for power is $P = I^2R$. This is the only formula that should be used to find power in AC systems.

**Power in an RL circuit.**

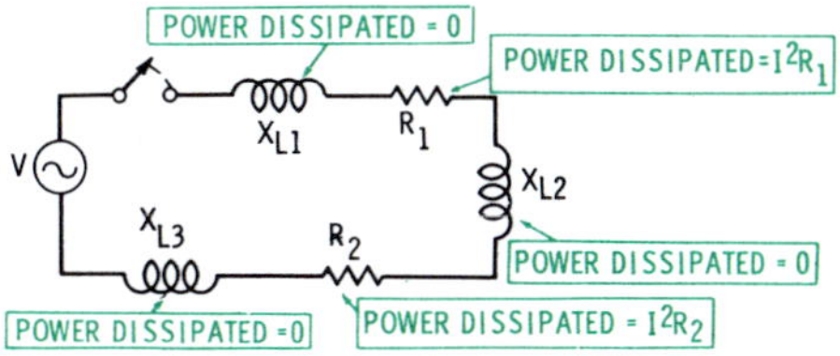


Go through all the steps necessary to find the power dissipated in the following circuit. (See Q24 through Q27.)

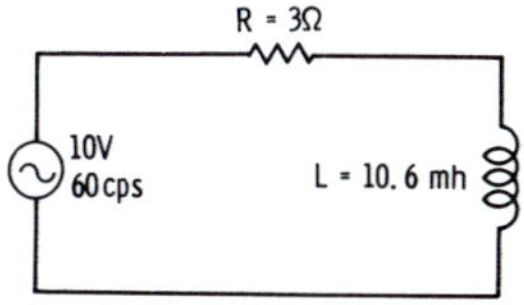


**Q24. What is the inductive reactance of the coil?**

**Q25. What is the impedance of the circuit?**

**Q26. How much current flows through the circuit?**

**Q27. How much power is dissipated in the circuit?**

**Q28. Here is a more complicated circuit. Notice how the current divides in the parallel parts. Kirchhoff's law is used to determine how the current divides. How much power is dissipated in this circuit?**

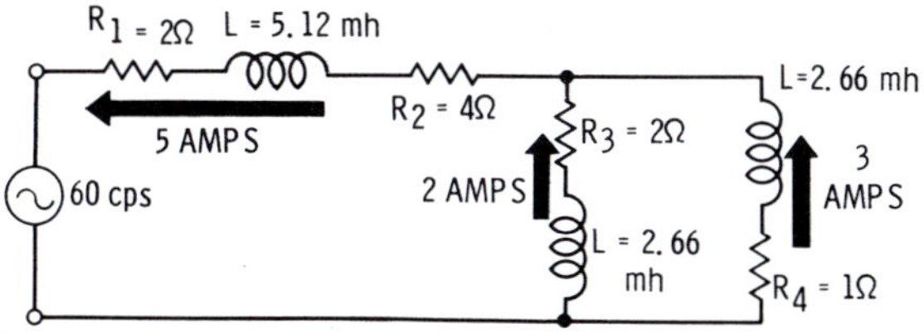


**Q29. Suppose a 4-henry filter choke and a 2,630-ohm resistor are wired in series across the terminals of a transformer that supplies 15 volts at 60 cps. Draw a schematic of the circuit.**

**Q30. How much current will flow in the circuit?**

**Q31. How much power will be dissipated by the resistor?**

**Your Answers Should Be:**

**A24.** $X_L = 2\pi fL =$ **4 ohms** (approx)

**A25.** $Z = 5\underline{/53^\circ}$ **ohms**

**A26.** $I = \frac{E}{Z} =$ **2 amperes**

**A27.** The power dissipated is: $P = I^2R =$ **12 watts**

**A28.** Since the current is known, the inductors can be disregarded. Use $P = I^2R$ to find the power dissipated in each resistor.

$$P_1 = 5 \times 5 \times 2 = 50 \text{ watts}$$
$$P_2 = 5 \times 5 \times 4 = 100 \text{ watts}$$
$$P_3 = 2 \times 2 \times 2 = 8 \text{ watts}$$
$$P_4 = 3 \times 3 \times 1 = 9 \text{ watts}$$

TOTAL **167 watts**

**A29.**

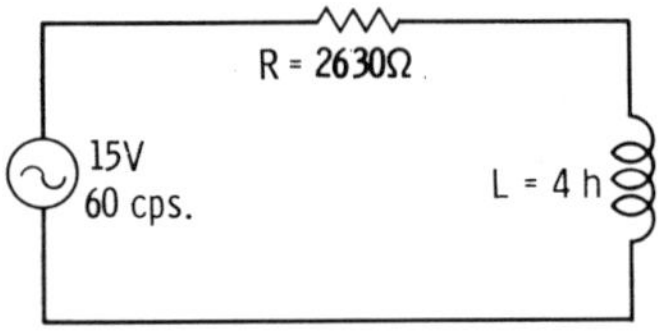


**A30.** $X_L = 1{,}507$ ohms $\quad I = \frac{E}{Z} = \frac{15}{3{,}000} =$ **5 ma** (approx.)

$R = 2{,}630$ ohms

$Z = 3{,}000 \underline{/30^\circ}$ ohms (approx.)

**A31.** $P = I^2R =$ **.0657 watt.**

## WHAT YOU HAVE LEARNED

1. Series and parallel inductive reactances can be combined in the same way as resistances.
2. L/R is the time constant of an RL circuit and indicates the time in seconds it takes a pulse to reach 63% of its maximum value.
3. Q is the ratio of $X_L$ to R in a coil.
4. The phase angle in an AC circuit is determined by the proportions of $X_L$ and R.
5. $X_L$ and R are added vectorially to find impedance.
6. Impedance may be expressed as $Z\underline{/\theta}$ or $R + jX_L$.

# 11

# The Effect of Capacitance

## *What You Will Learn*

**When you have finished this chapter, you will know how capacitance affects AC and pulses. You will learn how capacitance blocks DC but allows AC to pass more and more easily as the frequency increases. An explanation of how capacitance causes applied voltage to lag the current and how it distorts the voltage waveforms of pulses is given. You will become familiar with the units in which capacitance is measured and the factors that influence the value of a capacitor. You will be able to calculate capacitive reactance.**

### WHAT IS CAPACITANCE?

Capacitance is the property of an electrical circuit that **opposes a change in voltage.** Capacitance has the same reaction to voltage that inductance has to current. This means that if the voltage applied across a circuit is increased, capacitance will resist that change. If the voltage applied to a circuit is decreased, capacitance will oppose the decrease and try to maintain the original voltage.

In a DC circuit, capacitance has an effect only when voltage is first applied, and then again when it is removed. **Note that current cannot flow through a capacitance.** However, AC current **appears** to flow through a capacitance—you will learn how later. Since voltage is constantly changing in AC circuits, capacitance acts at all times to retard these changes in voltage.

A basic capacitor (sometimes called a condenser) is shown below. It consists of two conducting metal plates separated by a layer of air or other insulating material, such as paper, glass, mica, oil, etc. The insulating layer is called the **dielectric.**

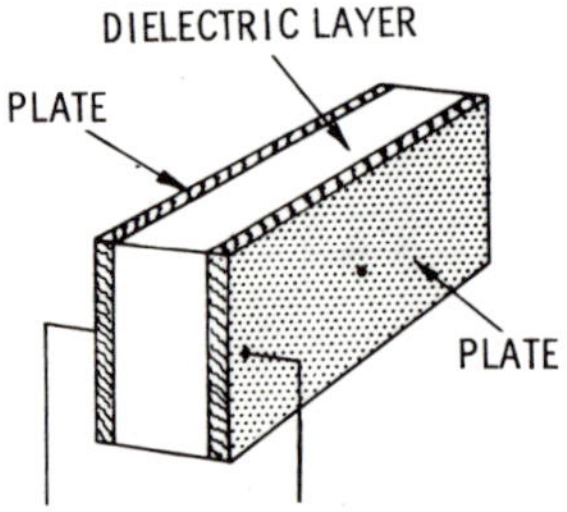


**A basic capacitor.**

All capacitors have these two plates and a separating layer. In practice, these are often stacked or even rolled into a compact form. Sometimes the dielectric is a paste or a liquid instead of a solid.

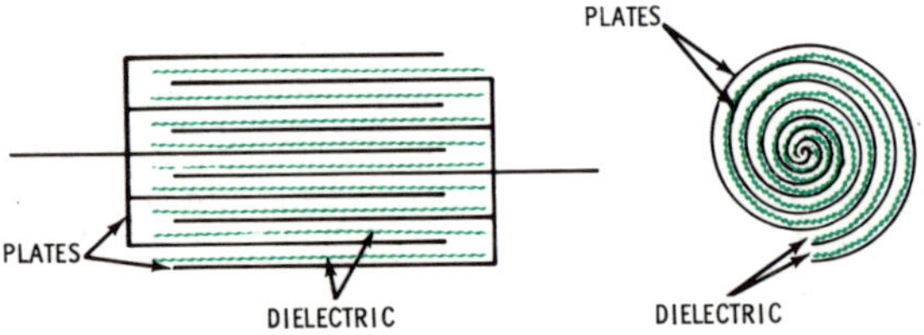


**Capacitor plates are often stacked or rolled to conserve space.**

This is the circuit symbol for a capacitor.

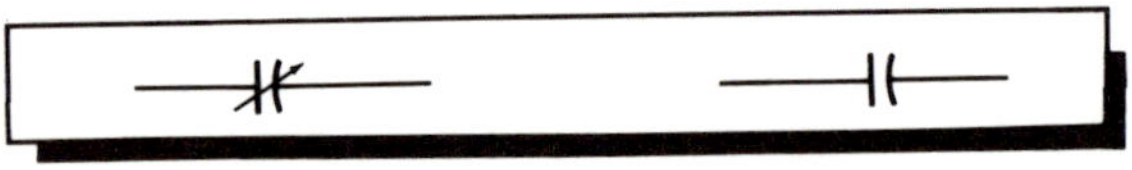

*Variable.* *Fixed.*

**Circuit symbols for a capacitor.**

When a capacitor is first connected to a battery, electrons from the negative terminal of the battery flow to the nearest capacitor plate and remain there. They can go no farther, since the opposite plate is separated from the first by an insulating layer. Electrons are moved from the opposite capacitor plate and flow into the positive terminal of the battery. After this initial movement of electrons, one plate is filled with all the electrons that the battery voltage can force into it, and the other plate loses the same number of electrons. This means that one plate has a negative charge and the other plate has an equal positive charge. No further current flows; the capacitor is "charged."

Positive and negative charges attract each other, so there will be a force between the plates of the capacitor. There is also a voltage between them that is equal to and opposes the voltage of the battery.

Because it takes a certain specific number of electrons to fill the negative plate, we say that the capacitor has a certain capacity, or **capacitance.**

You can see this happen if you take a capacitor (say 0.25 mfd, 600V) and connect the probes of an ohmmeter to the capacitor leads or terminals, using a very high-ohms range (R × 1 MEG). Notice that as soon as the connection is made, there is a sudden drop in the ohms reading as the battery in the ohmmeter provides current to charge the capacitor. Immediately following this decrease, the reading increases toward infinity. This shows there is some current at first (the charging current), but it quickly disappears and the resistance becomes infinitely large.

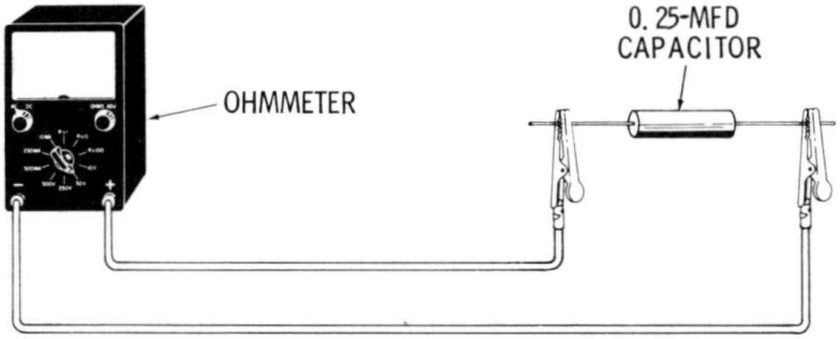


**Observing the charge of a capacitor.**

**Q1. Name two differences between capacitance and inductance.**

**Q2. Draw a circuit diagram of a capacitor connected across the terminals of a battery.**

**Q3. Explain what happens when you remove the battery from across a charged capacitor and place a shorting wire across the leads of the capacitor.**

**Q4. What would happen if you tried to repeat the above experiment without first discharging the capacitor?**

**Your Answers Should Be:**

A1. **Capacitance opposes a change in voltage** while inductance opposes a change in current. **Capacitance blocks DC** while inductance does not.

A2. Your circuit diagram should look like this.

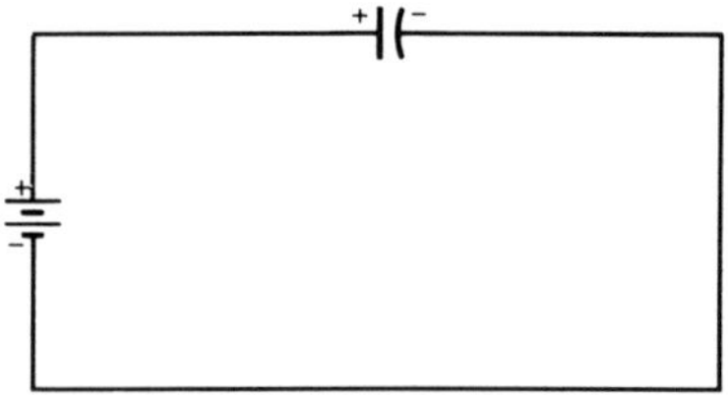

A3. The **electrons** from the negative plate **rush through the shorting wire to the positive plate** until both plates have the same number of electrons. The **voltage across the plates is then zero.**

A4. Once the capacitor has been charged, **there is no "kick"** (needle movement) when you connect the ohmmeter probes. The capacitor is already full and can take no more charge.

## CAPACITANCE MEASUREMENTS

The usual symbol for capacitance is **C.** Capacitance is measured in **farads.** The amount of capacitance in a capacitor is the quantity of electrical charges (in coulombs) which must be moved from one plate to the other in order to create a potential difference of 1 volt between the plates. The number of coulombs transferred is called the **charge.**

One farad is the capacitance in which a charge of 1 coulomb produces a difference of 1 volt between the plates. The larger the capacitance of a capacitor, the more charge it will hold with the same voltage applied across the plates.

Capacitance values are usually specified in microfarads (millionths of a farad, abbreviated mfd or **$\mu$f**) or in micromicrofarads. Micromicrofarads are also called picofarads (millionths of a microfarad, abbreviated **mmf, $\mu\mu$f,** or **pf**).

## HOW DOES CAPACITANCE AFFECT AC?

Although current cannot flow through a capacitor, an AC current appears to do just that. The reason lies in the nature of capacitance. If the voltage across the plates is continuously varied, the number of electrons on the plates varies.

Increasing the number of electrons on one plate of a capacitor repels electrons from the other plate. Decreasing the number of electrons on the first plate allows electrons to be attracted back to the other plate.

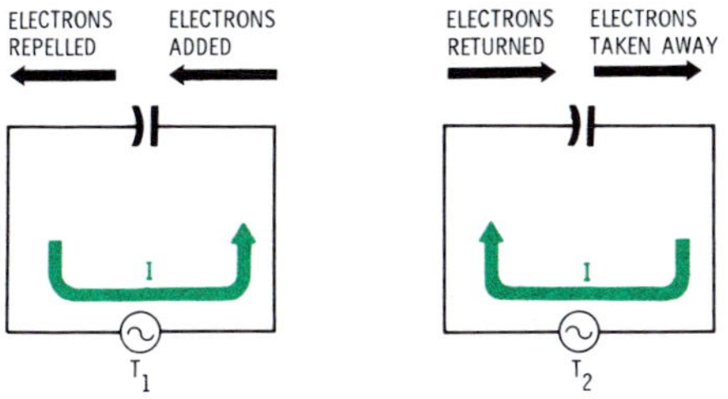


**AC can pass through a capacitor.**

An AC voltage can, in effect, get across the dielectric; since the voltage is alternating, it causes an AC current on the other side. In other words, **voltage changes** are transmitted across the gap.

If a capacitor has the same voltage as the applied voltage, no current will flow to or from it. If the applied voltage changes, the capacitor voltage will no longer equal the applied voltage. Current will flow trying to equalize the two.

In a circuit this means that if an AC sine-wave voltage is applied across a capacitor, an AC sine-wave current will appear on the opposite side, even though no electrons cross the dielectric layer.

**Q5. The capacitance of a capacitor is measured in ______.**

**Q6. A millionth of a farad is called a __________ and is abbreviated as ___ or __.**

**Q7. A ________________ is a millionth of a microfarad and is abbreviated ___, ___ or __.**

**Q8. Current will flow from one plate of a capacitor to the other plate only when _______ is changing.**

**Your Answers Should Be:**

**A5.** The capacitance of a capacitor is measured in **farads.**

**A6.** A millionth of a farad is called a **microfarad** and is abbreviated as **mfd** or **$\mu$f.**

**A7.** A **micromicrofarad** is a millionth of a microfarad and is abbreviated **mmf, $\mu\mu$f,** or **pf.**

**A8.** Current will flow through a capacitor only when **voltage** is changing.

## PHASE

Just as with inductance, current and voltage are not in phase in a capacitive circuit. The voltage lags the current (current leads the voltage) by 90°.

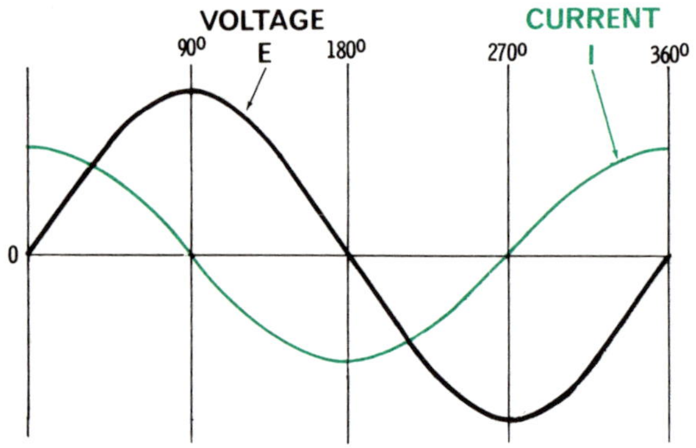


**Current leads voltage in a capacitor.**

At any instant, the current flowing into or out of a capacitor is proportional to the rate of change of the applied voltage. This can be seen in the illustration on the next page. The applied voltage is changing most rapidly at time A, the beginning of the sine-wave cycle. Therefore current flow is maximum. At time B the voltage across the capacitor has reached its peak and, for the moment, is not changing. Therefore current at this instant is zero. At time C, voltage across the capacitor again is changing quite rapidly (but in the negative direction), and so the current is at its negative peak. At time D, when the voltage reaches its negative peak and is momentarily not changing, the current waveform passes through zero once more.

If we trace the current from point to point along the voltage waveform, the result is a sine wave, but it is one that **leads the voltage by exactly 90°.** This shows that if the voltage across the capacitor is a continuous sine wave with a constant amplitude, the current through the capacitor circuit is a sine wave that is 90° ahead of the voltage.

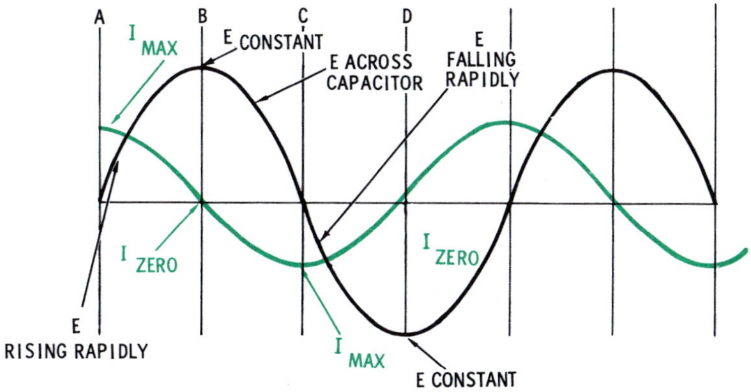


**Current is determined by the voltage change.**

Current and voltage vectors in a capacitive circuit are 90° out of phase. In this case the current vector is **ahead** of the voltage vector by 90°.

**The current vector leads the voltage vector.**

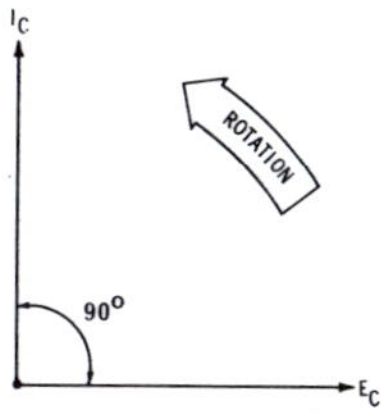


**Q9. When an AC voltage across a capacitor is maximum, AC current through the circuit is ----.**

**Q10. When an AC current through a capacitor circuit is maximum, AC voltage across the capacitor is ----.**

**Q11. Contrast the phase relationship of AC voltage and AC current in an inductor and in a capacitor.**

**Your Answers Should Be:**

**A9.** When an AC voltage across a capacitor is maximum, AC current through the circuit is **zero.**

**A10.** When an AC current through a capacitor circuit is maximum, AC voltage across the capacitor is **zero.**

**A11.** In an inductor, voltage **leads** current by 90°; in a capacitor, voltage **lags** current by 90°.

## FACTORS AFFECTING CAPACITANCE VALUE

The amount of electrical charge that can be stored in a capacitor (the number of electrons that can be placed on the plate) varies with the **area** of the plates. Consequently, capacitance varies directly with area—if the area is doubled, the capacitance is doubled. When the area is doubled or twice as many plates are connected in parallel, there is twice as much area to store electrons, and the capacitance is therefore twice as great.

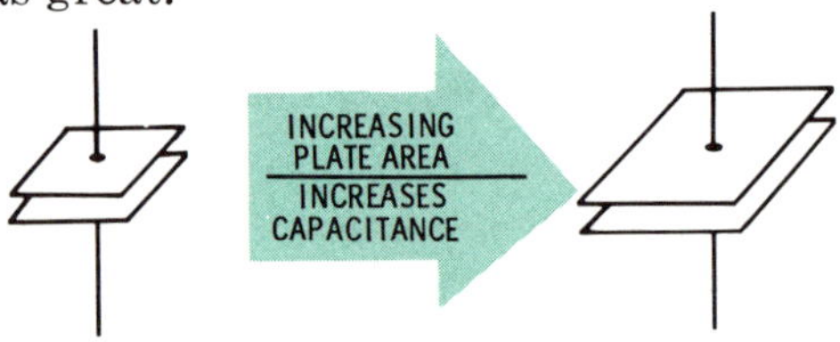


**Plate area affects capacity.**

Capacitance can also be increased by placing the plates closer together. When the plates are closer, the attraction

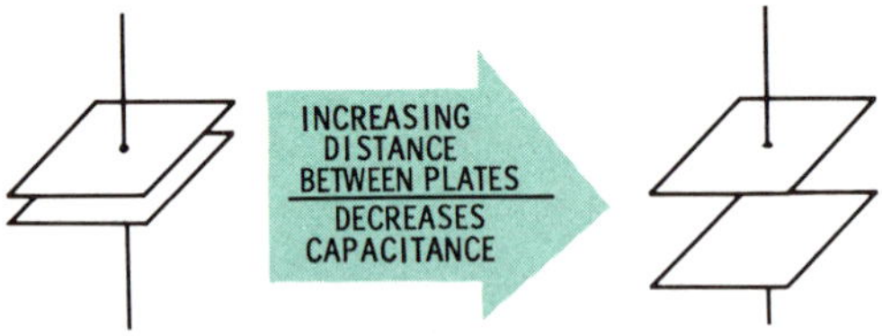


**Distance between plates affects capacity.**

between the negative charges on one side and the positive charges on the other side is greater, and thus more charge can be stored. It is, of course, necessary to keep the plates far enough apart so that the charge does not cross the gap.

Higher values of capacitance can be obtained by using an insulating material (dielectric) other than air. This allows the plates to be closer together without the charge crossing the gap.

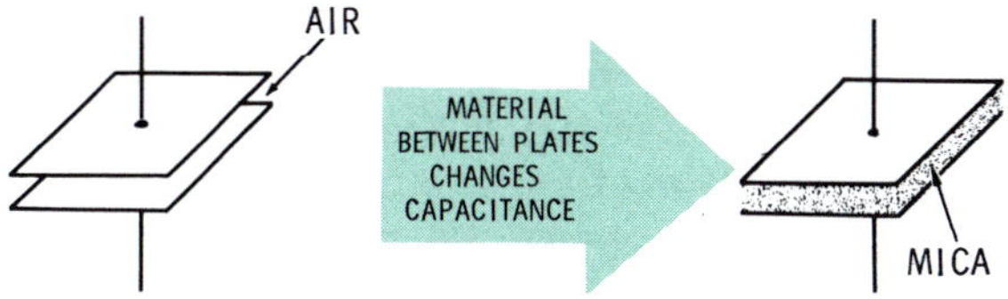


**Dielectric material affects capacitor.**

Dielectrics such as mica, glass, oil, and **Mylar** are a few of the materials that can sustain a high electric stress without breaking down. This property is called **dielectric constant.** The higher the dielectric constant is, the better the dielectric. Air has a dielectric constant of 1, glass about 5, and mica 2.5 to 6.6.

Besides allowing the plates to be placed closer together, a dielectric has another effect on capacitance. Dielectric material contains a large number of electrons and other carriers of electrical charge. Although electrons cannot flow as in a conductor, they are held rather loosely in the structure and can move slightly. The distortion of the structure of the dielectric, which is caused by charging the capacitor, has a large effect on the forces of attraction and repulsion that aid or oppose the flow of the electrons. This factor has a substantial effect on capacitance.

When materials such as mica or glass are used as the dielectric, the capacitors have a much higher value than the same size units with an air dielectric.

**Q12. If you had two capacitors of low value, how could you combine them to get a larger capacitance?**

**Q13. How does a mica capacitor differ from an air capacitor of the same physical size?**

**Q14. What are three factors that affect the capacitance of a capacitor?**

**Q15. A screw-type variable capacitor is made with an adjusting screw that is used to vary the distance between the capacitor plates. How would you increase its capacitance?**

**Your Answers Should Be:**

A12. **Combining two capacitors in parallel** would produce the same effect as a single capacitor with more plates, and therefore **would result in a higher capacitance.**

A13. A mica capacitor has a **higher capacitance** than an air capacitor of the same physical size.

A14. The capacitance of a capacitor depends on these three factors: the **area** of the plates, the **spacing** between the plates, and the nature of the **dielectric material.**

A15. Tightening the screw moves the plates closer together and **increases capacitance.** Loosening the screw **decreases the capacitance.**

## POWER

Just like inductance, capacitance consumes no power. During the sine-wave cycle, the capacitor takes energy out of the circuit and stores it in the form of an electric field during a quarter cycle and returns it to the circuit in the next quarter cycle. **Energy is borrowed, but it is always returned.**

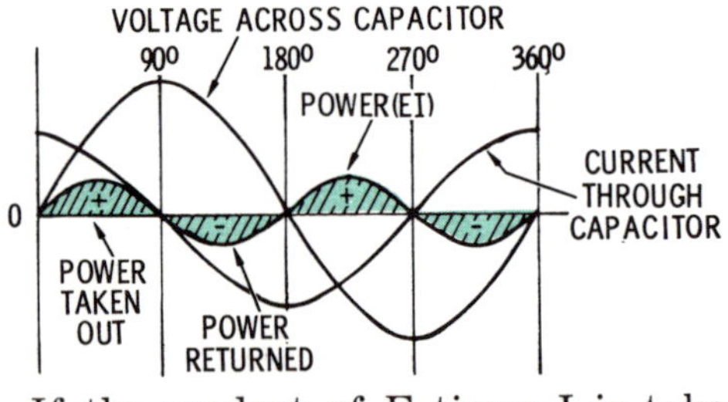


**Power in a capacitor.**

If the product of E times I is taken at every instant of the cycle, the power waveform will show that energy is taken out and returned in alternate quarter cycles.

To find the amount of energy (in **coulombs**) stored in a capacitor, multiply the capacity in farads by the applied voltage. In a circuit containing only pure capacitance, it makes no difference how long the voltage is applied—the same amount of energy will always be stored at a given voltage.

## CAPACITIVE REACTANCE

Like inductance, capacitance has a reactance—an opposition to the flow of AC. But capacitive reactance **decreases** as frequency increases.

Suppose a capacitor is connected in series with an alternating voltage source. There is no resistance in the circuit.

**A basic capacitive circuit.**

Because the circuit above contains no resistance, the voltage across the capacitor will be the same value as the source voltage at every instant.

When a capacitor is charged up to voltage E, it stores an amount of energy equal to the capacitance times the voltage. If the peak voltage of the AC source is E, the capacitor will have stored a particular amount of energy every time the voltage sine wave hits its peak, and again stores that amount whenever the voltage reaches it negative peak. The energy depends only on capacitance and peak voltage.

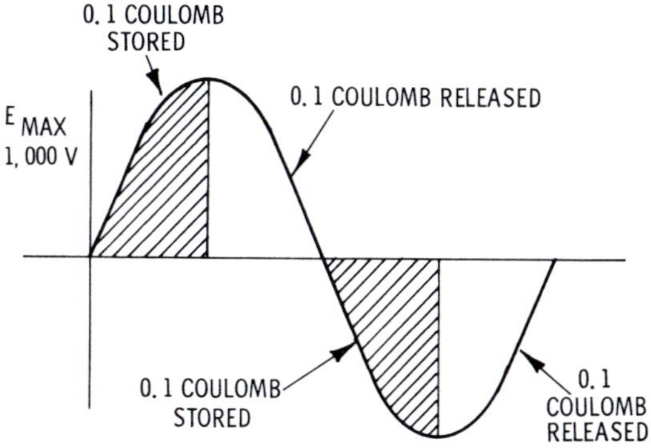


**Equal amounts of energy must flow in each cycle.**

**Q16. How much energy will be stored in a 100-mfd capacitor in the first quarter cycle of an applied AC voltage of 1,000 volts maximum?**

**Your Answer Should Be:**

**A16.** 1,000 volts × .0001 farad = **0.1 coulomb**

What happens when the frequency of the power source is doubled? If the peak voltage (E) is unchanged, the capacitor will charge every half cycle to the same amount as before. But it will have to do this twice as fast because the frequency is doubled. This means that the same amount of energy must flow into the capacitor in only half the time. And since the voltage is the same, we must have twice the current to supply this same amount of total energy.

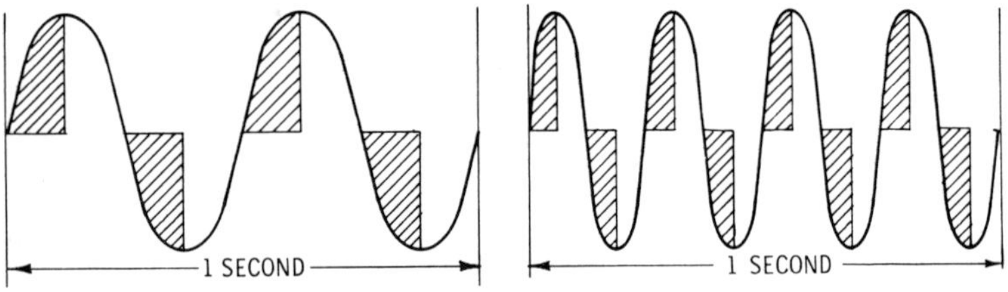


**A capacitor stores the same amount of energy each time it reaches $E_{MAX}$.**

What does this mean? The frequency was doubled, and this doubled the current flowing into the capacitor—yet, the input voltage remained the same. A pure capacitance lets twice as much current flow if the frequency is doubled.

**Capacitive reactance** is the opposition that pure capacitance offers to the flow of current. It is expressed in **ohms,** and its symbol is $\mathbf{X_C}$. Capacitive reactance depends on frequency. As the frequency increases, the rate of change of applied voltage increases, and the current flowing also increases. As the frequency is reduced, the rate of change of voltage goes down, and less current flows.

At this point you can more easily see why capacitor current leads the voltage across the capacitor. It is necessary for the capacitor to charge up to the given voltage, and this charging is done by the current. Hence, the charging current will reach its maximum value at the time the charging is going on at the greatest rate; that is, when the rate of change of voltage is the most rapid.

As the capacitor approaches full charge, the voltage rate of change slows down, and the current decreases. When the capacitor is fully charged and its voltage has reached maximum, there is no charging current flowing at all—the current has already dropped to zero at this time. A similar process occurs during discharging. At all times, current leads the voltage by 90°, or one quarter of the cycle. In a steady-state AC situation, when the applied voltage is a sine wave, both voltage and current will be sine waves.

Capacitive reactance depends on frequency. Since it lets more current flow as frequency increases, capacitive reactance must decrease as the frequency increases.

Capacitive reactance also depends on the size of the capacitance. As capacitance increases, more current must flow into the capacitor to charge it to the same voltage (since the amount of energy stored equals C × E). As a result, capacitive reactance decreases when capacitance increases.

The formula for capacitive reactance is:

$$X_C = \frac{1}{2\pi f C} \text{ ohms}$$

where,

f is the frequency in cps,
C is the capacitance in farads.

Capacitive reactance can be used in calculating current in a purely capacitive circuit by Ohm's law.

$$I = \frac{E}{X_C}$$

**Q17. What is $X_C$ if f = 6,000 cps and C = 200 mfd?**

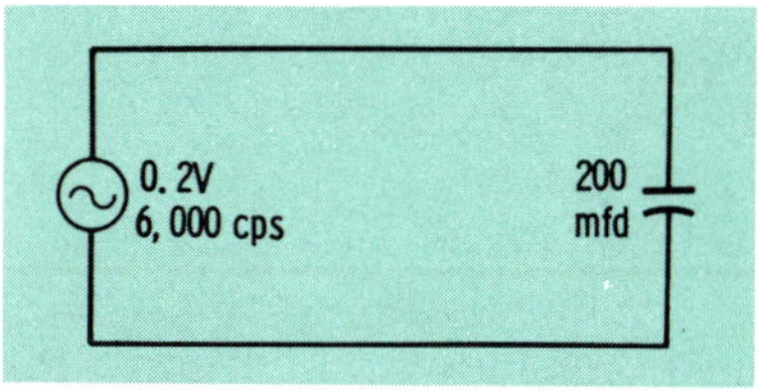


**Q18. What is the current in this circuit?**

**Q19. What would the current in the above circuit be if the input signal were 0.01 volt at 120 kc?**

**Your Answers Should Be:**

**A17.** $X_C = \frac{1}{2\pi fC} = \frac{1}{2 \times 3.14 \times 6{,}000 \times 200 \times 10^{-6}}$

$= \frac{1}{7.53} = \mathbf{0.133\ ohm}$

**A18.** $I = \frac{E}{X_C} = \frac{0.2}{0.133} = \mathbf{1.5\ amps}$

**A19.** $I = \frac{0.01}{0.0066} = \mathbf{1.52\ amps}$

## PULSE RESPONSE OF CAPACITANCE

When a sharp pulse, such as a square wave, is applied to a circuit containing capacitance, the capacitance opposes the sudden change of voltage. This results in a rounding off of the sudden voltage rise. Similarly, when the pulse voltage is suddenly decreased, the voltage across the capacitor does not decrease suddenly, but it trails off. Current is greatest when the change of voltage is greatest, so the current waveform will have a peak when the voltage rises suddenly, and another peak (but in the opposite direction) when it drops.

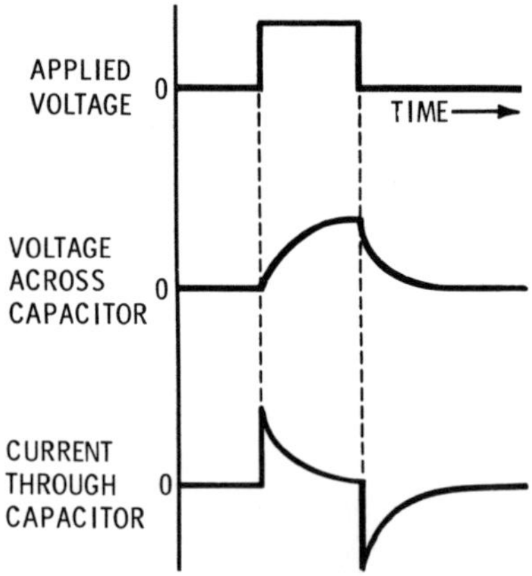


**Current is greater when voltage change is greatest.**

There is always some resistance in a circuit. By choosing the right value of capacitance and resistance, a circuit can be designed in which the voltage takes a predetermined length of time to reach a certain value. This type of circuit can thus provide a time delay.

## APPLICATION OF CAPACITANCE

You have already learned how inductance can be used in filters. Capacitance can also be used to block DC or to bypass AC. You have also learned that capacitance can be used for time delays. Another use for capacitance is to store energy. For example, capacitors are often used to store energy in photoflash units. In these photographic devices, a battery that produces a relatively small current can be used to gradually charge a capacitor. The capacitor releases a large store of energy very quickly when it is discharged through the flashbulb.

## STRAY CAPACITANCE

Capacitive reactance decreases as frequency increases. In communications, pulse, and radar work, where very high frequencies are used, **stray capacitance** can present quite a problem.

In a vacuum tube, an antenna, or a receiver chassis, there are always very small capacitances between adjacent conductors and between conductors and nearby objects which are meant to be isolated from each other. With the lower radio frequencies these capacitances are not important, but as the frequency increases, the capacitive reactances of these very small capacitances decrease. A decrease in reactance can actually cause leakage of the signal.

It is important to know, therefore, that at high frequencies, placement of wires and components is very important in order to keep the effects of stray capacitances to a minimum.

**Q20. How do the effects of capacitance on pulses differ from the effects of inductance?**

**Q21. Compare and contrast capacitive reactance and inductive reactance on these points:**

1. **Effect of an increase in frequency on reactance.**
2. **Effect of reactance on DC.**
3. **Effect of phase relations in AC.**

**Q22. What constant value appears in the formulas for both capacitive and inductive reactance?**

**Your Answers Should Be:**

A20. Capacitance **rounds off the voltage waveform and produces spikes in the current waveform.** Inductance **rounds off the current waveform and produces spikes in the voltage waveform.**

A21. 1. $X_C$ **decreases** as frequency increases, while $X_L$ **increases.**

2. $X_C$ **blocks DC,** while $X_L$ **passes DC.**

3. Capacitance causes current to **lead** the applied voltage, while inductance causes it to **lag.**

A22. $2\pi$ appears as a constant in both formulas.

## WHAT YOU HAVE LEARNED

1. Capacitance offers an opposition to a change in voltage.
2. A basic capacitor consists of metal plates separated by a dielectric.
3. A capacitor stores electrical energy in the form of an electric field as the capacitor charges, and releases this energy when it discharges.
4. Capacitance is a measure of the energy storage capacity of a capacitor. This capacity is measured in farads.
5. A capacitor blocks DC but allows AC to flow.
6. Pure capacitance in a circuit causes current to lead the applied voltage by 90°.
7. The amount of capacitance is determined by the area of the plates, the distance between them, and the dielectric material.
8. A capacitor stores energy and returns it to the circuit.
9. The opposition of capacitance to the flow of AC is called capacitive reactance.
10. The formula for capacitive reactance is:

$$X_C = \frac{1}{2\pi fC}$$

11. Capacitance rounds off the voltage waveform of a pulse.
12. Stray capacitance can cause signal leakage at high frequencies.

# 12

# RC Circuits

## *What You Will Learn*

**When you have finished this chapter you will know how to find the equivalent capacitance of combinations of capacitors in series and in parallel. You will be able to add capacitive reactance and resistance vectorially in order to find impedance. You will be able to analyze RC circuits and to explain how they affect various voltages and currents. You will know how to find RC time constants and how they are used.**

### A BASIC CAPACITIVE CIRCUIT

First, let's review what you have learned about capacitance by applying it to this basic capacitive circuit.

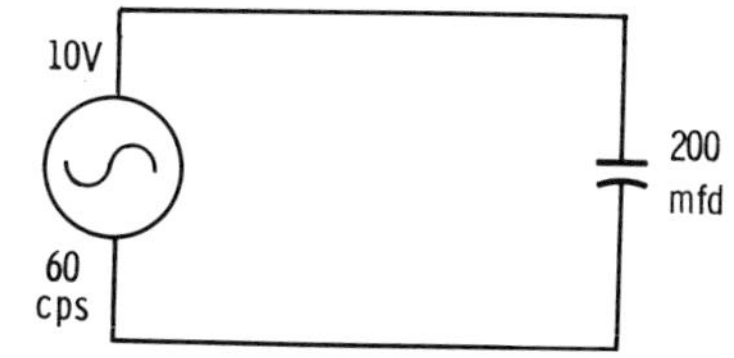


**A basic capacitive circuit.**

You have already learned that when a sine-wave AC voltage is applied, the current in a capacitor always leads the voltage by 90°. You have also learned that a capacitance consumes no power; all the energy it takes out of a circuit in one quarter cycle is returned in the next quarter cycle.

Both of the above statements are true, not only for a single capacitance, but also for any combination of capacitors. In fact, any circuit that contains only pure capacitances, no matter how many, can be simplified to include only one representative equivalent capacitance.

## CAPACITORS IN COMBINATION

As you know, resistors and inductors add in series. Two resistors or inductors in series have the same effect as a single, larger resistor or inductor.

*Resistors.*

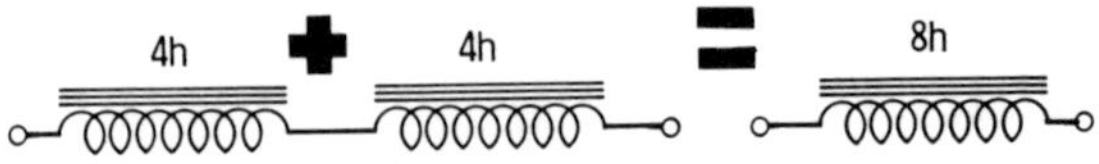


*Inductors.*

**Resistors and inductors add in series.**

### Capacitors in Parallel

Capacitors add in parallel. It is easy to understand why this is true if you remember that the more plates a capacitor has, the greater is its capacitance.

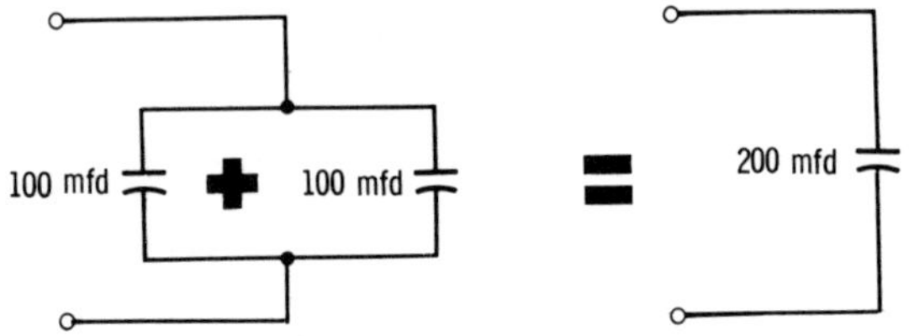


**The more plates a capacitor has, the greater is its capacity.**

If two capacitors are connected in parallel, you can find their equivalent capacitance by adding their values.

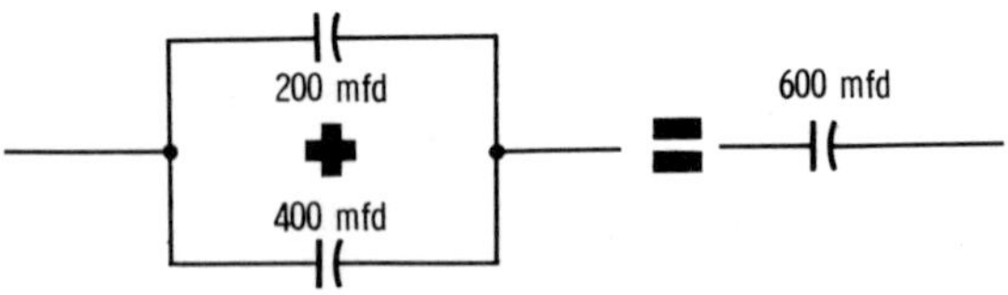


**Capacitors add in parallel.**

If a 200-mfd and a 400-mfd capacitor are connected in parallel, the equivalent capacitance of the combination is

200 mfd plus 400 mfd, or 600 mfd. This is also true with three, four, or any other number of capacitances in parallel.

**Capacitors in Series**

If two capacitors are connected in series, what is their equivalent capacitance? We cannot simply add $C_1$ and $C_2$ together. Instead, we use the relationship:

$$C_{eq} = \frac{C_1 \times C_2}{C_1 + C_2}$$

What is the total capacitance of 200 mfd and 400 mfd connected in series? Using the above relationship, the total capacitance is calculated to be:

$$C_{eq} = \frac{200 \times 400}{200 + 400} = 133 \text{ mfd}$$

Notice that this equivalent value is smaller than either of the two individual capacitor values.

**Capacitors in series.**

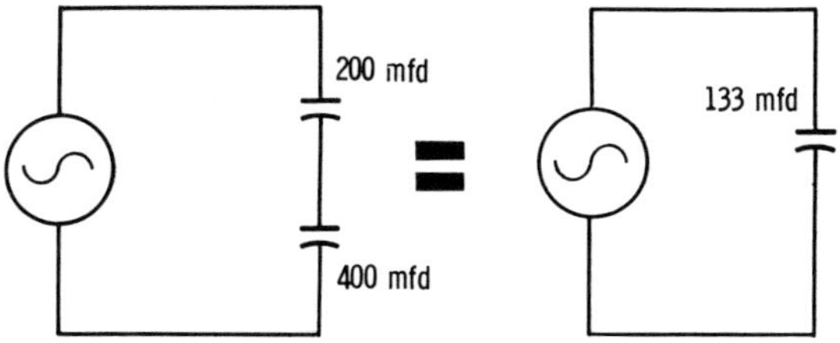


**Q1. What is the equivalent capacitance of the following circuit?**

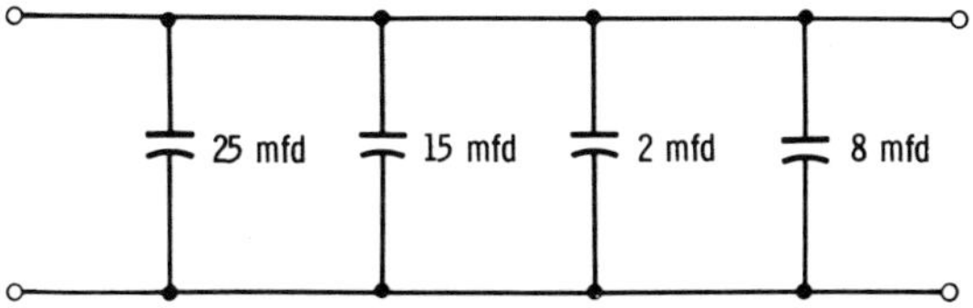


**Q2. How would you replace a 500-mfd capacitor if you had available only 200-mfd and 300-mfd capacitors?**

**Q3. Would you expect two capacitors in series to have greater or less capacitance than the same two capacitors in parallel?**

**Q4. What is the equivalent capacitance of two equal capacitors in series?**

**Your Answers Should Be:**

**A1.** $C_{eq}$ is 25 + 15 + 2 + 8 = **50 mfd.**

**A2.** Connect a 200-mfd and a 300-mfd capacitor **in parallel** to get an equivalent capacitance of 500 mfd.

**A3.** Two capacitors in series would have **less capacitance** than the same two capacitors in parallel.

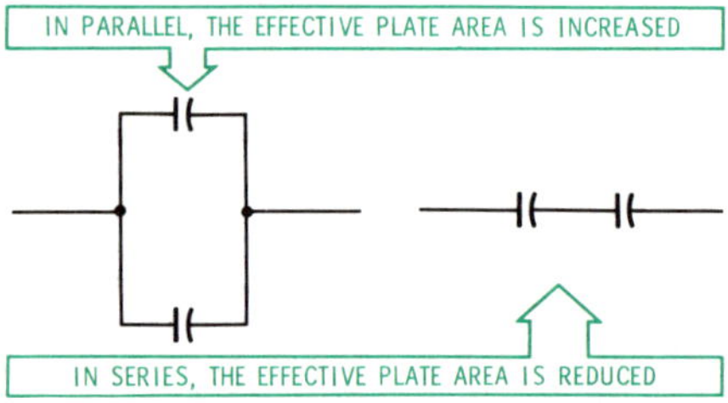


**A4.** The equivalent capacitance of two equal capacitors in series is always **one half** the capacitance of either one of them.

Complicated circuits can be simplified by analyzing them in steps as shown below.

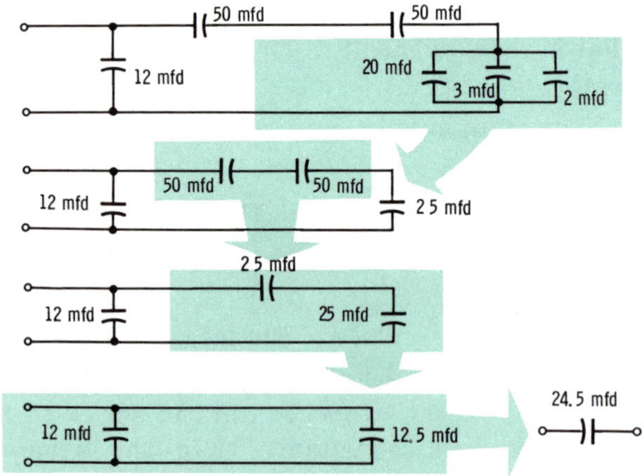


On the preceding pages, capacitance values were combined, not capacitive-reactance values. You must not confuse the two. To combine capacitive reactances, whether in series or in parallel, use the same rules that apply to resistance—add series values and combine parallel values by the formula:

$$X_{eq} = \frac{X_1 \times X_2}{X_1 + X_2}$$

Always check to make sure that all the capacitive-reactance values are computed for the same frequency—otherwise the result will be wrong. In order to calculate the reactance of a capacitive circuit, it is usually better to compute the equivalent capacitance first, and calculate the capacitive reactance of the equivalent capacitance.

**Q5. What is the equivalent capacitance of this circuit?**

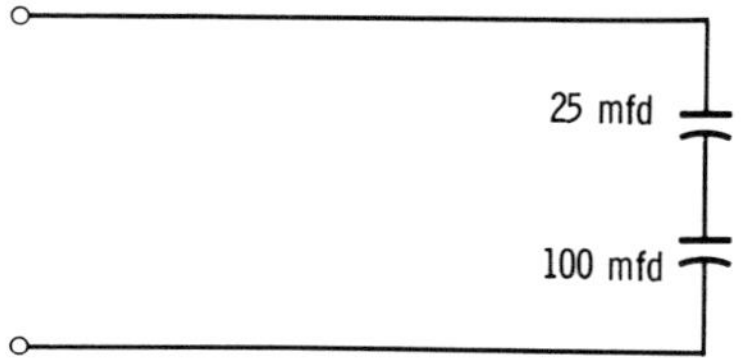


**Q6. What is the equivalent capacitance of this circuit?**

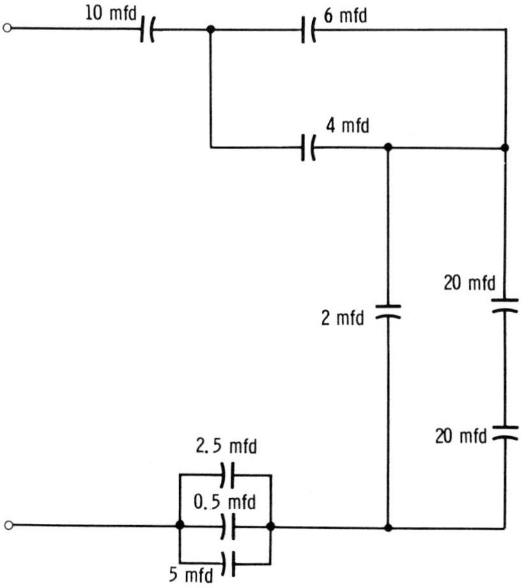

**Your Answers Should Be:**

**A5.** $C_{eq} = \frac{25 \times 100}{25 + 100} = \frac{2{,}500}{125} = \mathbf{20\ mfd}$

**A6.** You should have simplified the circuit to something like this:

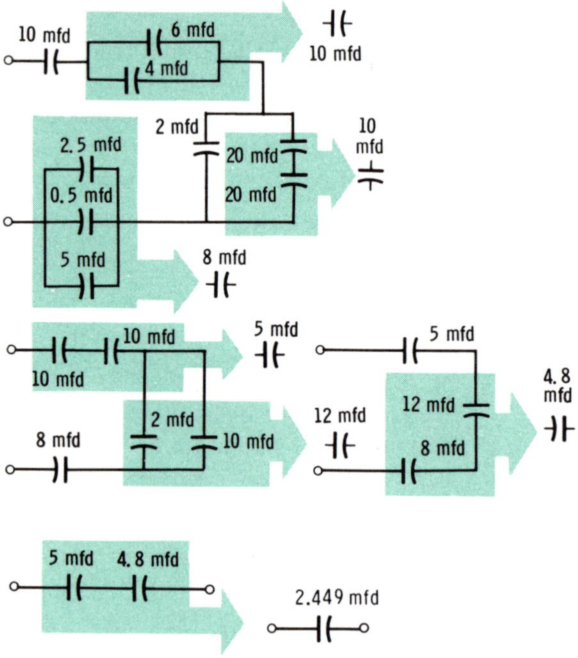


## RC CIRCUITS

Actually, an entirely pure capacitance does not exist. The leads of capacitors have some small value of resistance. Also, the dielectric layer is never quite perfect; it has some extremely high value of resistance. So if you wanted to be very accurate, you would represent these unwanted resistances by inserting their values in your circuit diagram, and treating them just as if they were actual resistors. For most practical purposes, however, you can disregard them.

Now let's see what happens if we put capacitors and resistors in the same circuit. As you already know, we cannot add resistance and capacitance, because they are two different quantities (resistance is measured in ohms, capacitance in farads). Instead, it is necessary to use capacitive reactance, which you learned about in the previous chapter. However, just as with inductance, to add resistance to capacitive reactance it must be remembered that resistive current is in phase with the voltage, while capacitive current leads the voltage by 90°. So the two cannot be added directly—they must be added vectorially.

**Capacitive reactance and resistance must be added as vectors.**

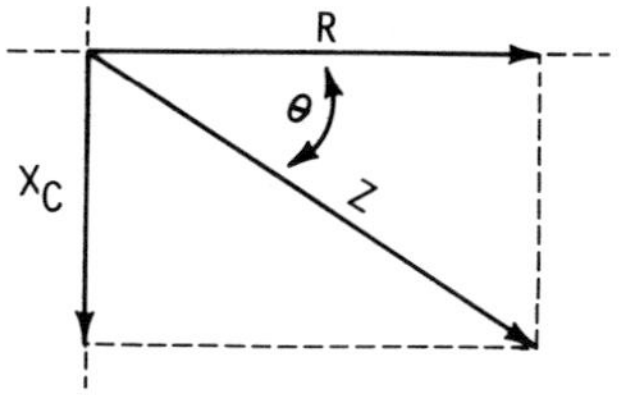


## IMPEDANCE

The capacitive-reactance vector is 90° **behind** the resistance vector. The resulting quantity, an **impedance,** is somewhere between the two, and its length (quantity) is the diagonal of the rectangle they form. This is **capacitive impedance,** which is different from inductive impedance because it lags the resistance vector. The way to write capacitive impedance is $R - jX_c$; the minus sign tells the story. All inductive impedances are represented by a + j, and all capacitive impedances by a — j in front of the X.

**Q7. Inductive impedance (leads, lags) resistance.**

**Q8. Would an impedance of 15 — j20 ohms be inductive or capacitive?**

**Q9.**

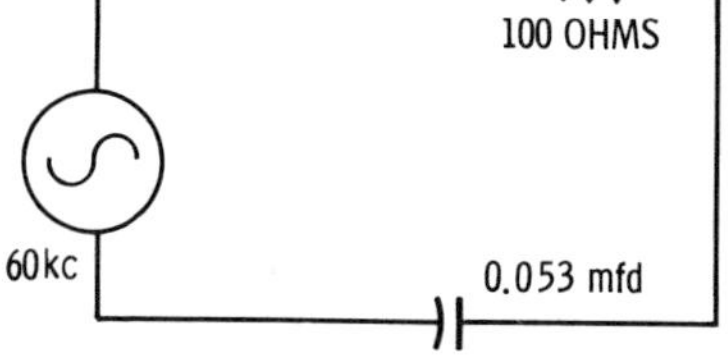


**Draw a vector diagram on graph paper to find the impedance in the circuit above. Measure the phase angle with a protractor.**

**Your Answers Should Be:**

**A7.** Inductive impedance **lags** resistance.

**A8.** An impedance of 15 — j20 ohms would be **capacitive.**

**A9. The capacitive reactance is 50 ohms.**
Your vector diagram should look like this.

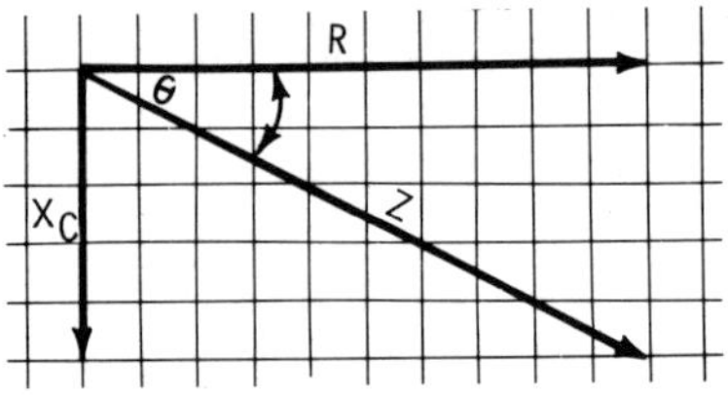


*Z = 100 — j 50 ohms*

*or 112 /—27° ohms*

If you have a table of trigonometric functions, you can get more accurate measurements of impedance, using the same formulas you used to find inductive impedance. Remember, these formulas apply to both inductive and capacitive impedance.

| | | |
|---|---|---|
| $R = Z \cos\theta$ | $X_c = R \tan\theta$ | $X_c = Z \sin\theta$ |
| $R = \dfrac{X_c}{\tan\theta}$ | $\dfrac{X_c}{R} = \tan\theta$ | $Z = \dfrac{R}{\cos\theta}$ |
| $\dfrac{R}{Z} = \cos\theta$ | $\dfrac{X_c}{Z} = \sin\theta$ | $Z = \dfrac{X_c}{\sin\theta}$ |

**Table of Trigonometric Functions**

| Angle | Sin | Cos | Tan | Angle | Sin | Cos | Tan |
|---|---|---|---|---|---|---|---|
| 5° | .087 | .996 | .087 | 50° | .766 | .643 | 1.192 |
| 10° | .174 | .985 | .176 | 55° | .819 | .574 | 1.428 |
| 15° | .259 | .966 | .268 | 60° | .866 | .500 | 1.732 |
| 20° | .342 | .940 | .364 | 65° | .906 | .423 | 2.144 |
| 25° | .423 | .906 | .466 | 70° | .940 | .342 | 2.747 |
| 30° | .500 | .866 | .577 | 75° | .966 | .259 | 3.732 |
| 35° | .574 | .819 | .700 | 80° | .985 | .174 | 5.671 |
| 40° | .643 | .766 | .839 | 85° | .996 | .087 | 11.430 |
| 45° | .707 | .707 | 1.000 | 90° | 1.000 | .000 | ∞ |

Use the partial table of trigonometric functions on the opposite page to solve the following problems.

**Q10. Find Z and $\theta$ for these values of $X_C$ and R.**

**(a) $X_C = 3$ ohms, R = 4 ohms.**

**(b) $X_C = 4$ ohms, R = 3 ohms.**

**(c) $X_C = 10$ ohms, R = 10 ohms.**

**(d) $X_C = 87$ ohms, R = 50 ohms.**

**Q11. Find $X_C$ and R for these values of $\theta$ and Z.**

**(a) $\theta = 45°$, Z = 10 ohms.**

**(b) $\theta = 37°$, Z = 0.5 ohms.**

**(c) $\theta = 15°$, Z = 1,000 ohms.**

**Q12. If $\theta$ is 60°, will R be larger or smaller than $X_C$?**

**Q13. If $\theta$ is 37°, will R be larger or smaller than $X_C$?**

**Q14. Will a circuit with an impedance of 100 ohms and a phase angle of 75° dissipate more or less power than a circuit with the same impedance and a phase angle of 45°?**

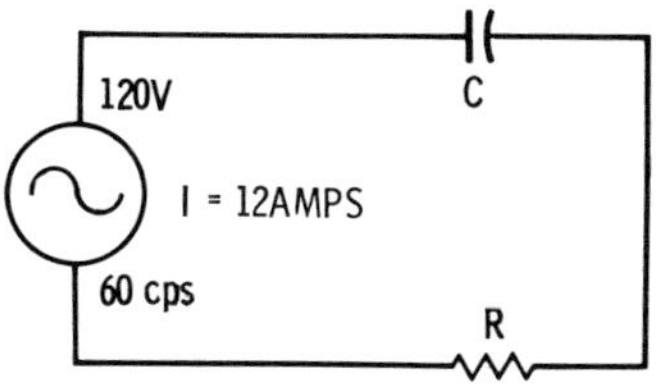


**Suppose you know the input voltage, the frequency, and the current flowing in the above circuit. Also, suppose you know that the phase angle of the circuit is 60°.**

**Q15. What is the DC resistance of this circuit?**

**Q16. What is the impedance of the circuit?**

**Q17. What is the capacitive reactance of the circuit?**

**Q18. What is the resistance of the circuit?**

**Q19. How much power is dissipated in the circuit?**

**Q20. What will happen to the capacitive reactance of the circuit if the frequency of the input is decreased?**

**Q21. What will happen to the phase angle of the circuit if the frequency of the input is increased?**

**Your Answers Should Be:**

**A10.** (a) **$Z = 5$ ohms, $\theta = 37°$**

(b) **$Z = 5$ ohms, $\theta = 53°$**

(c) **$Z = 14.14$ ohms, $\theta = 45°$**

(d) **$Z = 100$ ohms, $\theta = 60°$**

**A11.** (a) **$X_C = 7.07$ ohms, $R = 7.07$ ohms.**

(b) **$X_C = 0.3$ ohm, $R = 0.4$ ohm.**

(c) **$X_C = 259$ ohms, $R = 966$ ohms.**

**A12.** If $\theta$ is 60°, **R will be less than $X_C$.**

**A13.** If $\theta$ is 37°, **R will be more than $X_C$.**

**A14.** An impedance of 100 ohms will contain less resistance and **dissipate less power** at a phase angle of 75° than at a phase angle of 45°.

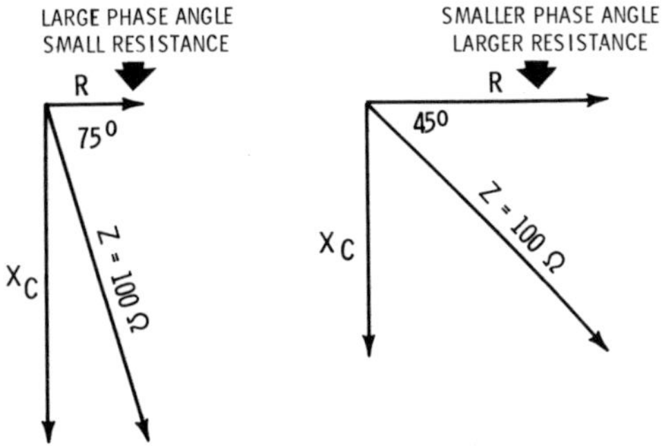


**A15.** The DC resistance will be **infinite** because the capacitor blocks DC.

**A16.** The impedance can be found by using the formula $Z = \frac{E}{I}$. **$Z = 10$ ohms.**

**A17.** $X_C = Z \sin \theta =$ **8.66 ohms.**

**A18.** $R = Z \cos \theta =$ **5 ohms.**

**A19.** $P = I^2R =$ **720 watts.**

**A20.** $X_C$ will **increase** if the frequency decreases.

**A21.** The phase angle will **decrease** if the frequency increases, because $X_C$ will decrease.

## RC TIME CONSTANT

The ratio between R and C has an important effect on the characteristics of a circuit. The way this ratio affects AC voltages and currents is indicated by a **time constant,** in much the same way that the effects of combined inductance and resistance were indicated.

What happens if you apply a pulse, such as a square wave, to a series RC circuit? The capacitor will oppose the sudden

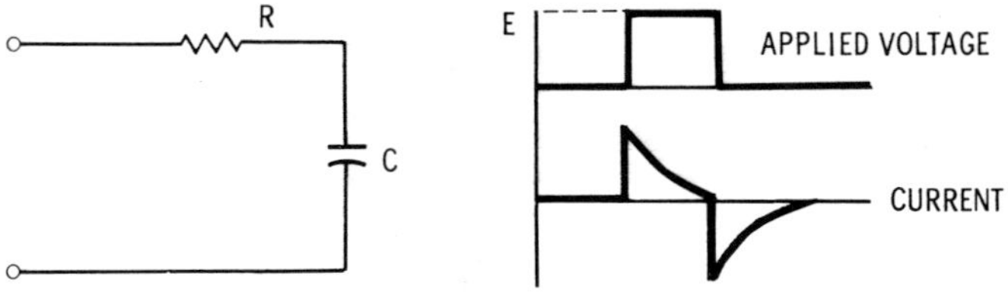


**A square pulse through an RC circuit.**

change of voltage and will gradually **charge** to source voltage E. The rate of charge (the initial current that will flow) is limited by resistance R. In fact, the initial current will be $I = E/R$ and will gradually decrease to zero as the voltage **builds up** across the capacitor.

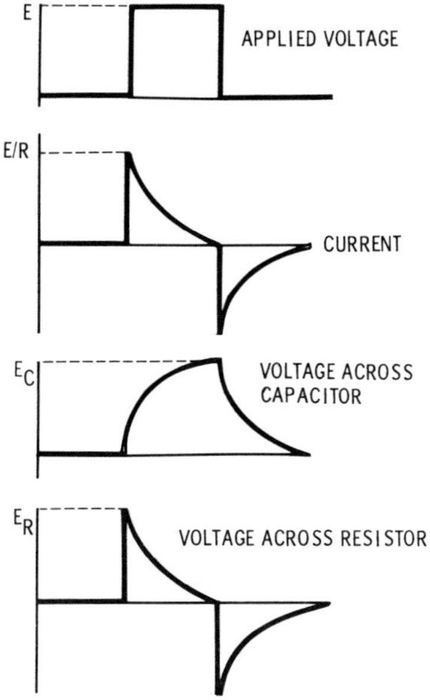

The voltage across the capacitor will start at zero and will build up smoothly until it equals source voltage E. The voltage across the resistor will, at any instant, equal the difference between the source voltage and the voltage across the capacitor; it will also be a spike as shown.

The rate of charging—the steepness of the capacitor voltage curve—depends on how much current the resistor will allow to flow. The higher the resistance, the less the current flow and the slower the charging rate will be.

This fact is expressed in numbers by the **time constant** of the circuit. The time constant of a series RC circuit is simply **R × C,** where R is in ohms, C is in farads, and the time constant is in seconds. **RC** is the time it takes the capacitor to charge to 63.2% of the source voltage.

For example, if R is 10,000 ohms and C is 10 microfarads, RC is $10{,}000 \times \frac{10}{1{,}000{,}000} = 0.1$ second.

When a capacitor has been charged, it actually contains a certain amount of stored energy. The stored energy is C times E coulombs, where C is the capacitance in farads and E is the voltage to which the capacitor is charged.

If a charged capacitor is connected in a circuit, its stored energy is released into the circuit. An example of this is a battery-capacitor, photographic-flash circuit.

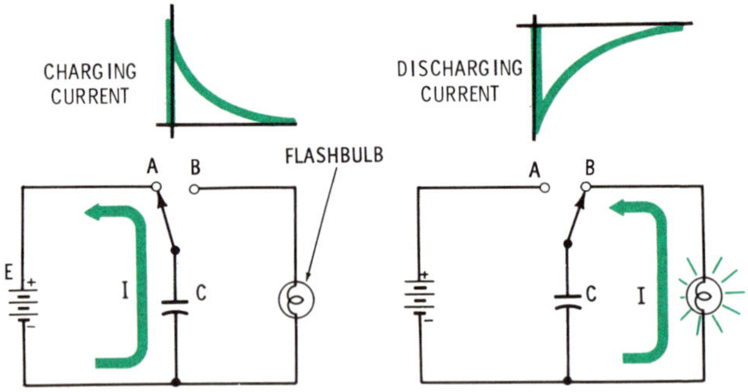


**Charging from battery.** **Discharging through bulb.**

Capacitor C is charged up to the battery voltage by throwing the switch to position A. The rate of charge depends on the resistance in the battery circuit (wire resistance and internal resistance of the battery). When a flash is desired, the switch is moved over to position B. The capacitor, which is charged to full battery voltage (E), has no opposing voltage in the new circuit, and its discharge is limited only by the resistance of the flash bulb. The stored energy flows through the flash bulb and, in doing so, fires the bulb. The discharging current follows the curve shown in the figure; the speed of discharge again depends on the time constant of the circuit.

An RC series circuit can be used as a timing device. It is, in fact, often used this way (e.g., in television receivers).

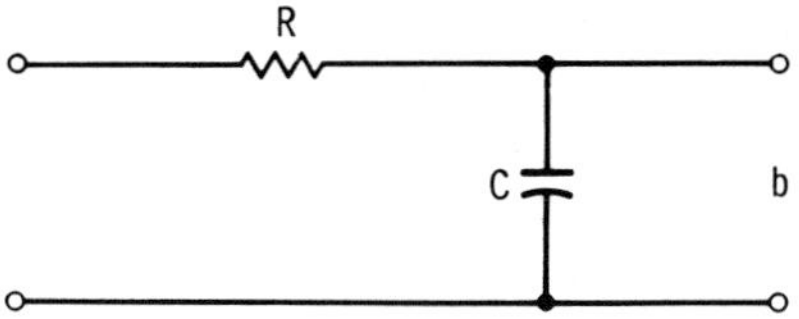


**A timing circuit.**

In the above circuit, the length of time it will take for the capacitor voltage to rise to some given value can be calculated. If some device, which will be triggered only when this given value of voltage is reached, is connected across the output terminals **b**, the device, such as a gas diode, will be triggered after a predictable time delay from the time the input voltage (E) is applied.

The length of the time delay can be controlled by varying either the resistance or the capacitance. However, if the amount of energy stored is important, the delay can only be varied by changing the capacitance.

**Q22. What is the time constant of the following circuit?**

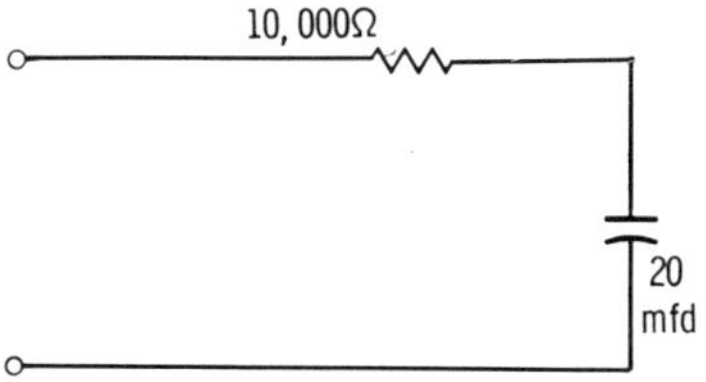


**Q23. How long will it take the circuit above to charge to 63.2 volts if a 100-volt input is continuously applied?**

**Q24. What will the time constant be if R is 1 megohm and C is 50 mfd?**

**Your Answers Should Be:**

A22. RC = 10,000 ohms × 0.00002 farad = **0.2 second**

A23. It will take **0.2 second** for the circuit to charge to 63.2 volts.

A24. $RC = \frac{50}{1{,}000{,}000} \times 1{,}000{,}000 =$ **50 seconds**

## WHAT YOU HAVE LEARNED

1. Capacitances add in parallel. They combine in series according to the formula:

$$C_{eq} = \frac{C_1 \times C_2}{C_1 + C_2}$$

2. Capacitive reactances add in series and combine in parallel exactly like resistances and inductances.
3. A circuit containing a complicated combination of capacitors can be converted to a simple equivalent circuit by a series of steps.
4. Capacitive reactance and resistance add vectorially.
5. The capacitive-reactance vector is in the opposite direction from the inductive-reactance vector and is represented in the impedance formula by —j.
6. Capacitive impedance can be calculated using trigonometric functions.
7. The time constant of a series RC circuit is found by multiplying R times C.
8. RC circuits can be used to store energy and/or to provide a time delay.
9. RC is the time in seconds it takes the capacitor to charge to 63.2% of the source voltage.

# 13

# RLC Circuits

## *What You Will Learn*

**You will now learn how to combine the two reactive effects (inductive and capacitive) with and without resistance. A number of important special circuits, called tuned circuits, depend on the RLC combination. When you have finished this chapter, you will know how to calculate their effects in circuits having inputs of various frequencies. You will also learn a general rule about phase and power in AC circuits, as well as how the power factor of a circuit affects the amount of power dissipated. You will be able to analyze the effects of RLC circuits on pulse inputs in terms of frequency response. You will know how to use a universal time-constant chart to analyze the reaction that RC and RL circuits have to step voltages and square waves.**

### RLC IMPEDANCE

When vector diagrams are used to find the impedance and phase angle (as in the previous chapters), $+jX_c$ is always drawn upward, while $-jX_c$ is always drawn downward. This leads to the idea that inductance and capacitance provide opposite reactions.

What happens if a circuit contains both inductance and capacitance in series? The two reactances cannot be just arithmetically added to find the total reactance. $+jX_L$ and $-jX_c$ tend to offset each other, and the total effect is their difference.

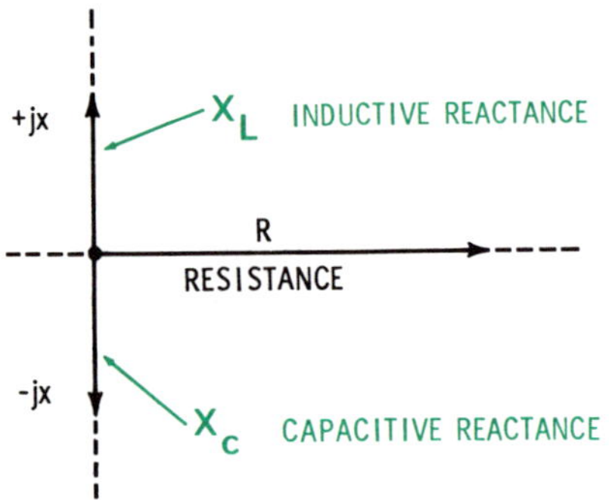


**The vectors for inductive and capacitive reactance are opposite.**

This difference is in the direction of the greater of the two reactances. So, if a circuit contains a capacitor, the reactance of which is —j50 ohms, and an inductor, the reactance of which is +j100 ohms, the net result is equivalent to an inductive reactance of +j50 ohms.

If a resistor is connected in series with an LC circuit, the impedance of the circuit will simply be the resultant reactance (whether inductive or capacitive) in series with the resistor.

A series circuit containing L and C behaves either as a capacitor or as an inductor, whichever of the two components has the greater reactance at the operating frequency.

## RESONANCE

A special case arises when the capacitive reactance and the inductive reactance are equal. When this condition exists, the reactances cancel each other and the circuit appears to be purely resistive. This can happen at only one frequency, however, for each particular set of inductive and capacitive values. At a low frequency, the inductive reactance is low and the capacitive reactance is high. The circuit, therefore, behaves as a capacitance. If the frequency of the applied voltage is gradually increased, the inductive reactance will gradually increase and the capacitive reactance will gradually decrease. At some point the two reactances become equal, and thus cancel. This point is called the

**resonant frequency** of the circuit. If the frequency is increased further, the inductive reactance becomes greater than the capacitive reactance, and the circuit will behave as an inductor.

Every L and C combination has one, and only one, resonant frequency. It is the frequency at which the inductive and capacitive reactances are equal.

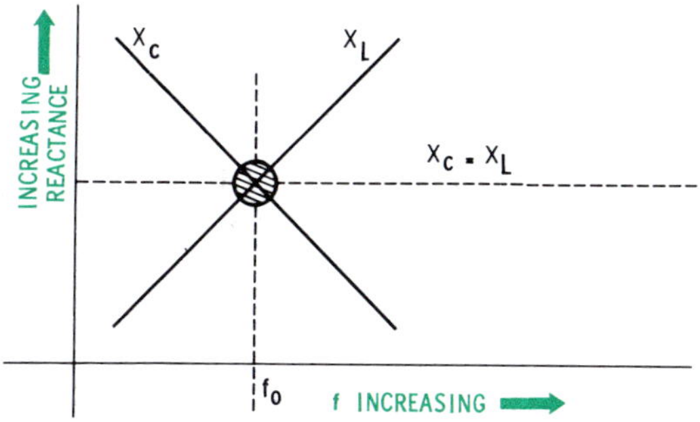


**Every LC combination has one resonant frequency.**

**Q1. Capacitance and inductance both store energy. Capacitance stores energy in a(an) _________ field. Inductance stores energy in a(an) _________ field.**

**Q2. How do capacitance and inductance affect a DC input?**

**Q3. What are the formulas for finding capacitive and inductive reactances?**

**Q4. How does an increase in the frequency of the input affect capacitive reactance? Inductive reactance?**

**Q5. Capacitive impedance is expressed in component form as R — jX. Inductive impedance is expressed as ______.**

**Q6. What is the total reactance of a circuit that has an $X_L$ of 100 ohms and an $X_C$ of 25 ohms? Is the total reactance capacitive or inductive?**

**Q7. The condition existing when capacitive reactance is equal to inductive reactance is known as __________.**

**Q8. Any circuit containing inductance and capacitance has only one _________ frequency.**

**Your Answers Should Be:**

A1. Capacitance stores energy in an **electric** field. Inductance stores energy in a **magnetic** field.

A2. **Capacitance blocks DC. Inductance offers no opposition to DC.**

A3. $X_C = \frac{1}{2\pi fC}$; $X_L = 2\pi fL$

A4. An increase in frequency results in a **decreased capacitive reactance,** and an **increased inductive reactance.**

A5. Inductive impedance is expressed as $\mathbf{R + jX}$.

A6. **75 ohms of inductive reactance.**

A7. The condition existing when capacitive reactance is equal to inductive reactance is known as **resonance.**

A8. Any circuit containing inductance and capacitance has only one **resonant** frequency.

### Resonant Frequency Calculation

The resonant frequency ($f_o$) formula is derived as follows:

$$2\pi f_o L = \frac{1}{2\pi f_o C}$$

Multiplying by $f_o$,

$$2\pi f_o^2 L = \frac{1}{2\pi C}$$

Dividing by $2\pi L$,

$$f_o^2 = \frac{1}{4\pi^2 LC}$$

Taking the square root of both sides,

$$f_o = \frac{1}{2\pi\sqrt{LC}}$$

Now see what actually happens in a series resonant circuit. Current I, which is in phase with the applied AC voltage, flows through all three components—L, C, and R. During the first quarter cycle (of each sine wave) the inductance is returning energy **to** the circuit and the capacitance is taking energy **from** the circuit at the same rate.

During the second quarter cycle, the situation is reversed—the capacitor is returning energy, and the inductor is taking it out. This sequence occurs during each cycle.

The voltage across the capacitance is equal and opposite to the voltage across the coil at all times, and the two cancel. One voltage ($E_C$) is 90° behind the current and the other voltage ($E_L$) is 90° ahead. No power is consumed in the L and C elements—only the resistor consumes power.

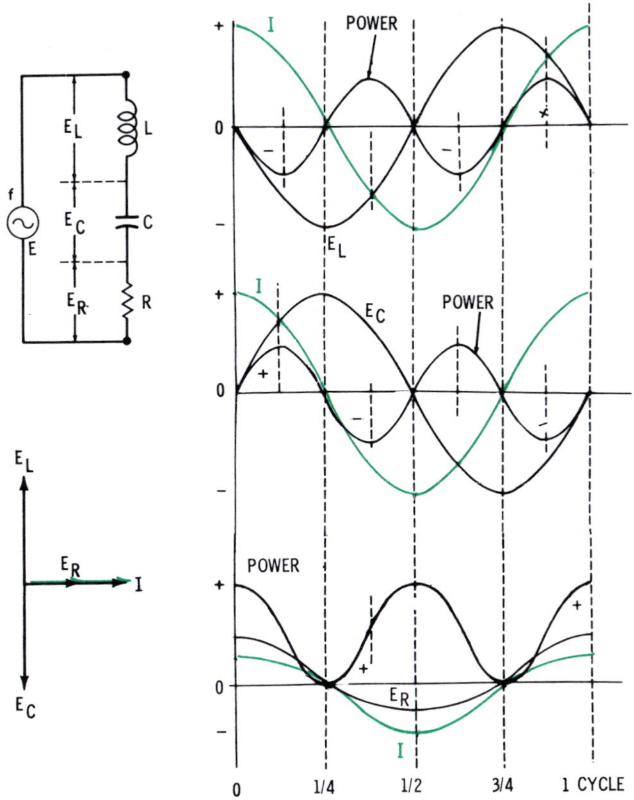


**Series resonance.**

**Q9. What is the resonant frequency of a circuit containing 2 henrys in series with 2 mfd?**

**Q10. If a 100-ohm resistor is placed in series with the two components of Question 9, what happens to the resonant frequency of the circuit?**

**Your Answers Should Be:**

**A9.** $f_o = \dfrac{1}{6.28 \times \sqrt{2 \times 0.000002}} = \mathbf{79.6\ cps}$

**A10.** If a 100-ohm resistor is placed in series with the two components of Question 9, the resonant frequency of the circuit will **remain the same.**

## Q of a Resonant Circuit

At resonance, the voltage across the capacitor and across the inductor is greater than at any other frequency. The effective current in the circuit is also higher at the resonant frequency than it is below or above resonance.

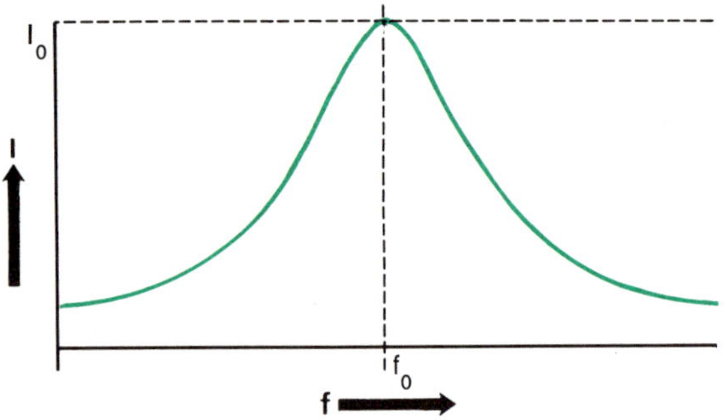


**Current in a series circuit is maximum at the resonant frequency.**

The quality of a resonant circuit can be measured by the Q factor. **Q** is the ratio of the energy stored in the capacitor and inductor divided by the energy dissipated in the resistor.

The amount of reactive opposition to current flow at a given frequency is not affected by the Q of the circuit. The resistive opposition, however, does vary according to the Q. This means that the shape of the resonance curve depends on this factor. If the frequency is changed from $f_o$ to a frequency where the reactance is low, and if the Q is high (resistance is only a few ohms), the total impedance will be halved. If the Q is low (resistance is high), the total impedance will be increased by only a small amount, and the current decrease will be very small.

The Q determines the exact shape of the resonance curve of a circuit. For example, if the resonant frequency is multiplied by $\frac{1}{Q}$, and the frequency of the input is changed from the resonant frequency by this amount, the current will be 0.707 times the resonant current. If the frequency is changed by $\frac{1}{2Q}$ times the resonant frequency, the current will be 0.447 times the resonant current.

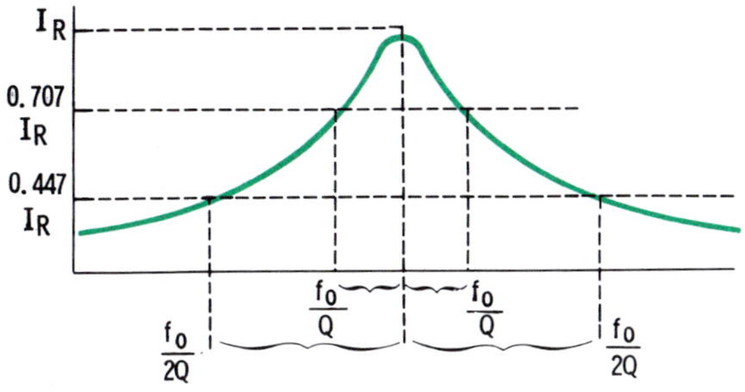


**The Q determines the shape of the resonance curve.**

**Q11. Both the capacitor and the inductor in a resonant circuit store energy. Do both store energy at the same time?**

**Q12. Do they both store the same maximum amount of energy?**

**Q13. At what frequency would you measure the Q of a circuit?**

**Q14. What is the Q of a circuit whose resonant frequency is 1,000 cps, inductance is 0.5 henry, and resistance is 10 ohms?**

**Q15. If two resonant circuits are identical except that one has a greater R (and therefore a lower Q) than the other, which one will pass a greater effective current at a given voltage?**

**Q16. Draw a curve representing current for various frequencies in a low-Q, series resonant circuit.**

**Your Answers Should Be:**

**A11. No.** The inductor stores energy when the capacitor is releasing it, and vice versa.

**A12. Yes.** When maximum energy is stored in the inductor, the capacitor has stored no energy.

**A13.** At its **resonant frequency.**

**A14.** $Q = \frac{X_L}{R} = \frac{2\pi fL}{R} = \frac{3{,}140}{10} = \mathbf{314}$

**A15. The one with the higher Q** will pass the greater current.

**A16.** Your curve should look something like this:

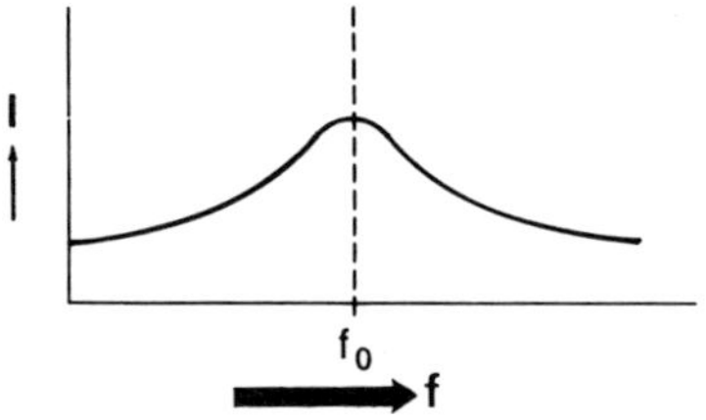


## APPLICATIONS

**Frequency-selective** properties of series resonant circuits are useful in applications where it is desired to pass one particular frequency with more ease than others. The circuit can act as a filter.

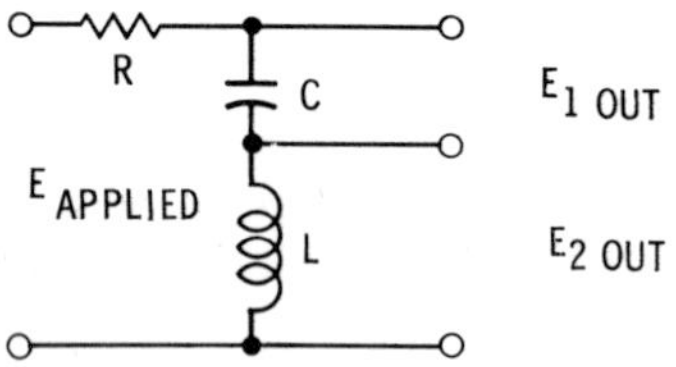


**A bandpass filter.**

If the voltage across either L or C is used for the output, the voltage will be much greater for signals having the resonant frequency than for signals above or below this frequency. Such a circuit is called a **bandpass filter.**

The width of the bandpass depends on the circuit Q—the higher the Q, the sharper the resonance curve, and the narrower the bandpass.

In a radio-frequency circuit a high-Q tuned circuit can be used to select the desired station and reject all others. In a power supply, a circuit using fairly large L and C values may be used to reject undesired frequencies.

## PARALLEL RESONANT CIRCUITS

A parallel resonant circuit is made up of inductance and each of the two branches shows reactance. The capacitive losses are usually associated with the coil rather than with the capacitor, the resistance is usually shown as being in series with the inductance.

**A parallel resonant circuit.**

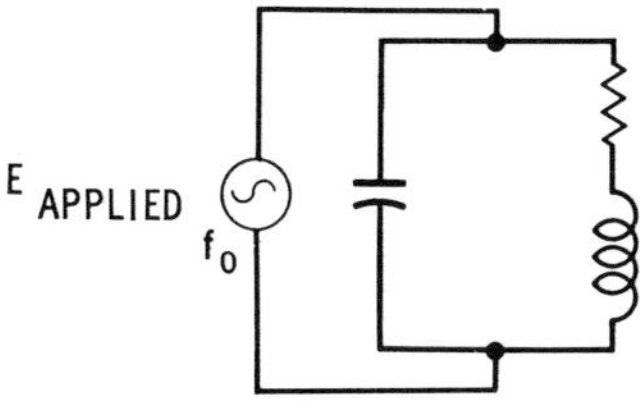


**Q17. What is the phase relationship between the applied voltage and the output voltage across the capacitor in the circuit on the opposite page?**

**Q18. What is the phase relationship between the applied voltage and the output voltage across the inductor?**

**Q19. What is the phase relationship between the applied voltage and the current through the circuit?**

**Q20. Analyze what will happen in the circuit on this page. Assume that R is negligible. What effect will C have on the phase of the current through the C branch?**

**Q21. What effect will L have on the phase of the current through the L branch?**

**Q22. Draw a sketch of the applied voltage sine wave and then of the inductive and capacitive currents to show their phase relationships.**

**Your Answers Should Be:**

**A17. Output voltage** across the capacitor **lags** the applied voltage **by 90°.**

**A18. Output voltage** across the inductor **leads** the applied voltage **by 90°.**

**A19.** Current is **in phase** with the applied voltage.

**A20.** In the C branch, current will **lead** the applied voltage **by 90°.**

**A21.** In the L branch, current will **lag** the applied voltage **by 90°** (if R is negligible).

**A22.** Your sketch should look like this:

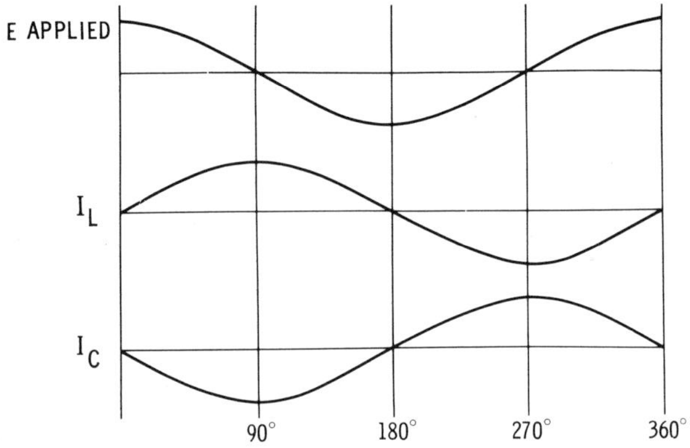


## Parallel RLC Circuits

When an AC voltage is applied to a parallel RLC circuit, each of the two branches shows reactance. The capacitive reactance in the capacitor branch is high at low frequencies, and decreases as the frequency increases. Similarly, the inductive reactance of the inductor branch is low at low frequencies, and increases as the frequency increases.

The capacitor has a high reactance and the inductor a low reactance at frequencies below resonance. Consequently, most of the current flows through the inductive branch and lags the applied voltage. Similarly, if the frequency is above resonance, most of the current will flow in the capacitive branch and will lead the applied voltage.

At some particular frequency the two reactances in a parallel resonant circuit are exactly equal. Since there is an AC voltage applied across each branch, both kinds of current are present—an inductive current in the inductive branch and a capacitive current in the capacitive branch. At resonance the two currents are equal.

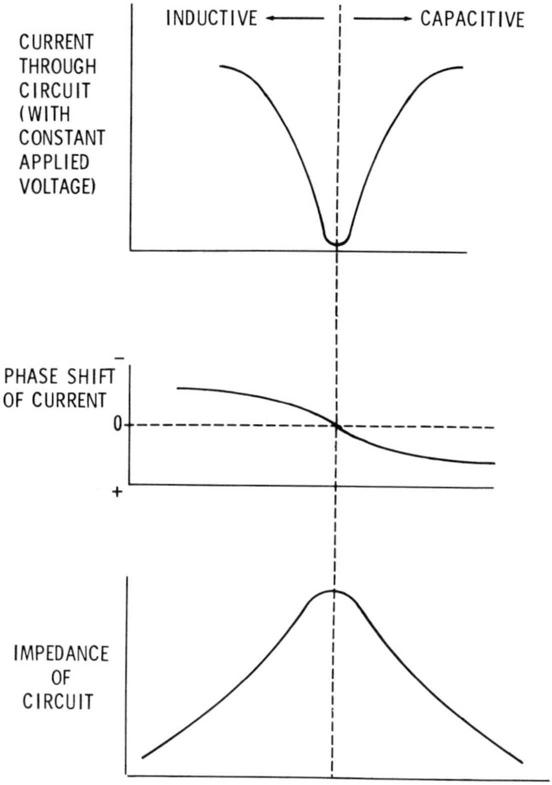


But, because one of the currents **leads** the applied voltage by 90° and the other **lags** the voltage by 90°, the two currents are 180° out of phase with each other. This means that they cancel (add up to zero).

The applied voltage was kept constant as the frequency was varied. Since current is minimum through the circuit at resonance, a parallel circuit has a higher impedance at the resonant frequency than at any other.

Now, let's look at what happens inside the loop formed by the inductance and capacitance.

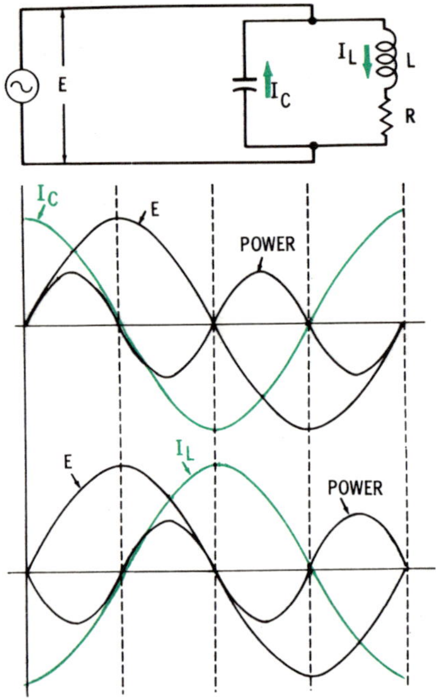


**Current, voltage, and power relationship in a parallel resonant circuit.**

The two large currents—inductive and capacitive—still flow, but only inside the loop. Energy alternately flows from capacitor to inductor and back again, twice each cycle. The capacitor alternately charges and discharges, first in one direction and then in the other. The inductive magnetic field alternately builds up and collapses, changing polarity twice each cycle. But all this flow back and forth is contained in the loop, and none appears in the external circuit. The outside circuit only has to replenish the energy lost in any resistance the inductor has, and this constitutes the entire external current.

The Q of the circuit, just as in the series resonant circuit, is the inductive reactance at resonance divided by the resistance of the inductor $(\frac{X_L}{R})$. In a parallel resonant circuit the loop current flowing between inductor and capacitor is Q times the external (resistive) current.

A parallel resonant circuit is, in a way, the opposite of a series resonant circuit. A series circuit has low impedance at resonance (maximum current); a parallel circuit has high impedance (minimum current). The total impedance will be greater as $X_L$ and $X_C$ become greater relative to resistance. It will decrease as R increases and draws more current. The impedance of a parallel circuit at its resonant frequency can be found by this formula:

$$Z_o = \frac{X_L X_C}{R}$$

If you use the formulas for $X_L$ and $X_C$, you can also develop another formula.

$$X_L\, X_C = 2\pi fL \times \frac{1}{2\pi fC} = \frac{2\pi fL}{2\pi fC} = \frac{L}{C}$$

Substituting in the above impedance formula,

$$Z_o = \frac{X_L X_C}{R} = \frac{L}{CR}$$

Another useful formula can be developed if you notice that $X_L/R$ is the Q of the circuit. This means that the impedance at resonance is simply $X_C$ (or $X_L$ since they are equal) times the Q of the circuit.

$$Z_o = X_L Q = 2\pi fLQ$$

The impedance curve for a parallel resonant circuit is the same shape as the current curve for a series resonant circuit. Its shape depends on the Q of the circuit in the same way.

Parallel tuned circuits are used in receivers, transmitters, and similar equipment. For instance, the five-tube radio receiver has parallel tuned circuits in its IF amplifiers. Their function is to select certain frequencies and reject others.

**Q23. What is the impedance at resonance of a parallel resonant circuit consisting of a 1-henry inductor with 1 ohm of DC resistance, and a 1-mfd capacitor?**

**Q24. If the resonant frequency of a circuit is 1,000 cps, L is 1 henry, and the Q of the circuit is 80, what is the impedance of the circuit at resonance?**

**Your Answers Should Be:**

**A23.** $Z_o = \frac{L}{CR} = \frac{1}{0.000001 \times 1} =$ **1 megohm**

**A24.** $Z_o = 2\pi fLQ = 6.28 \times 1{,}000 \times 1 \times 80 =$ **502,400 ohms**

## POWER IN RLC CIRCUITS

To calculate the power dissipated in a circuit containing only resistance, either $P = I^2R$ or $P = EI$ can be used. Either will yield the same result. However, $P = EI$ applies only to resistive circuits (circuits in which the voltage and current are in phase).

In a parallel circuit that contains resistance and either capacitance or inductance, $P = EI$ gives the **apparent power.** This, however, is not the true power—apparent power is always larger. The reason is that the overall current in a reactive circuit is not in phase with the voltage. The total current is actually the vector sum of the resistive current (which is in phase with voltage) and the reactive current (which leads or lags by 90°). To use $P = EI$, multiply the voltage by only that portion of the current that is in phase—the resistive current. In order to calculate the true consumed power, the resistive and reactive currents must be taken separately.

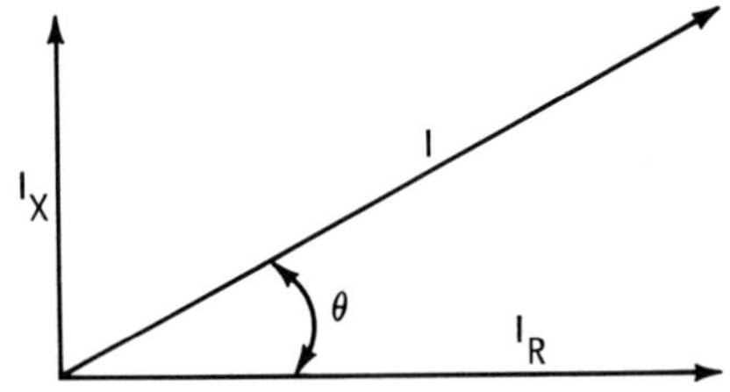


**The overall current is the vector sum of the reactive current and the resistive current.**

The overall current is the vector sum of the reactive current and the resistive current. This overall current leads or lags the applied voltage by an angle $\theta$.

The true power dissipated in a circuit is found by multiplying the apparent power by cos $\theta$. True power divided by apparent power equals cos $\theta$ and is called the **power factor** of the circuit.

$$\frac{EI \cos \theta}{EI} = \cos \theta$$

There are several things to consider about power factor. The more reactance there is in a circuit (compared to the resistance present), the more out of phase the current will be with the voltage, and the smaller the power factor will be.

It is possible to represent the total current flowing as two separate component currents at right angles to each other. The resistive current produces power (does the work). The reactive current merely flows in and out of the capacitance or inductance as it charges and discharges, but it does no work.

Even though the reactive current produces no power in the circuit, the wires must be large enough to carry it. Power lines are designed to carry the total current, not just the resistive component.

A low power factor sometimes leads to problems. For example, when a plant that uses induction motors draws a large reactive current (has a low power factor), equipment must be provided to supply it with more current than it actually consumes. The large inductive component can be canceled in such cases by placing large capacitors in series with the load. The capacitors draw reactive current 180° out of phase with the inductive current, and thus the capacitors cancel the inductive component. This is called "power factor improvement" and can save considerable money in large power installations. Another method sometimes employed by power companies to improve a low power factor is to persuade the consumer to use synchronous instead of induction-type motors. This results in an overall saving to the power company because less power need be generated and supplied to the consumer.

**Q25. Is a high or a low power factor desirable in electrical circuits used to transmit power?**

**Q26. How can inductive reactance be canceled to increase the power factor in an inductive circuit?**

**Your Answers Should Be:**

**A25.** A **high** power factor.

**A26.** Inductive reactance can be canceled by **capacitive reactance.**

## PULSES IN RLC CIRCUITS

Pulses are a special type of AC, even though they often have a DC component as well, and are very important in modern electronics. Radar systems are based on pulses of RF energy reflected from targets. Digital computers, counters, and other data-processing circuits employ pulses. Pulses are also used in telemetry and remote control to switch circuits on and off. Trains of pulses are used to transmit information between satellites, spaceships, and the earth.

All these applications create a need for circuits that can generate, amplify, send, receive, count, recognize, and/or process pulses. Because pulses are composed of a combination of many sine-wave frequencies, pulse-handling circuits have very severe requirements placed on them.

### Frequency Response

All pulse waveforms are made up of a combination of sine waves.

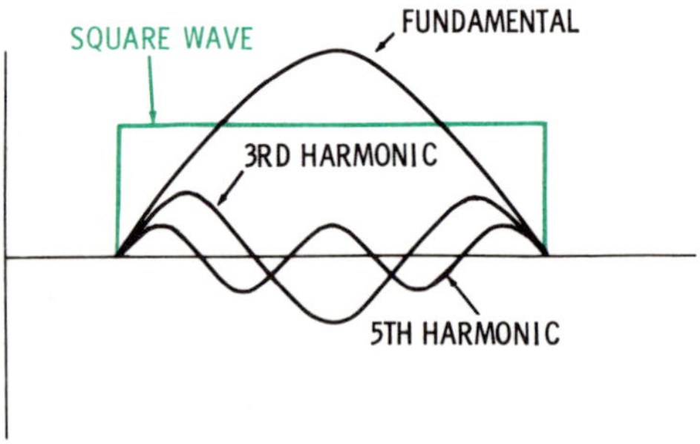


**A square wave is composed of sine waves.**

The lowest frequency contained in a train of rectangular pulses is called the fundamental frequency. This frequency has a period equal to that of the square wave. All the other frequencies contained in the square wave are **harmonics,** or multiples, of the fundamental frequency. In the case of a square or rectangular wave, only the **odd harmonics** are included. The next higher frequency is three times the fundamental (the third harmonic), then five times the fundamental (the fifth harmonic), etc. A perfectly square waveform contains an infinite number of odd harmonics, from the fundamental up.

To pass or amplify a square wave without distortion, a circuit must be able to pass all the frequencies contained in the wave. In practice, an infinitely high frequency response is not possible, nor is it necessary. This less-than-perfect response is because all circuits have some frequency-sensitive characteristics due to such factors as stray capacitance, the inability of vacuum tubes or transistors to amplify signals above certain limits, etc. All of these frequency-sensitive characteristics result in the corners of the square wave being rounded off.

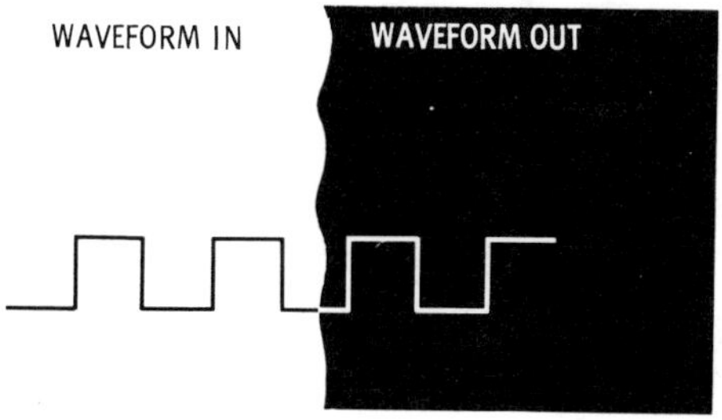


**The circuit responds the same way to all frequencies above the fundamental.**

**Q27. In a pulse circuit, inductance opposes a change in ________.**

**Q28. In a pulse circuit, capacitance opposes a change in ________.**

**Q29. What sine-wave frequency contributes most of the amplitude of the flat-top, or straight-line, portion of a square wave?**

**Q30. What sine-wave frequencies are responsible for the steep leading and trailing edges of a square wave?**

**Your Answers Should Be:**

A27. In a pulse circuit, inductance opposes a change in **current.**

A28. In a pulse circuit, capacitance opposes a change in **voltage.**

A29. The **fundamental** frequency.

A30. The **higher** harmonic frequencies.

## Pulse Circuit Applications

If a circuit has poor low-frequency response, there will be a visible sag in the level portion of the output waveform.

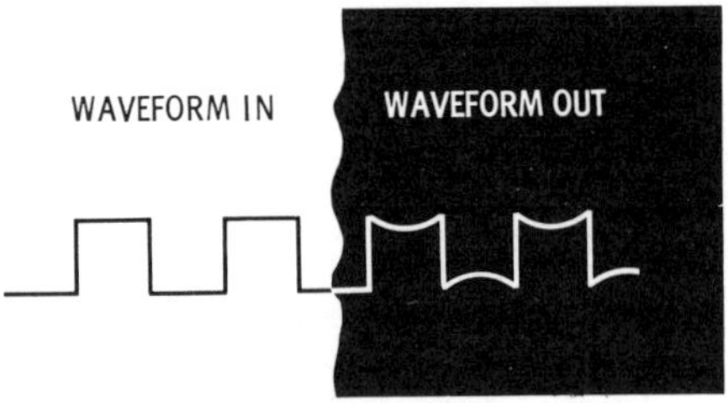


**Poor low-frequency response.**

If the circuit has good low-frequency response but poor high-frequency response, the corners of the output waveform will be rounded off.

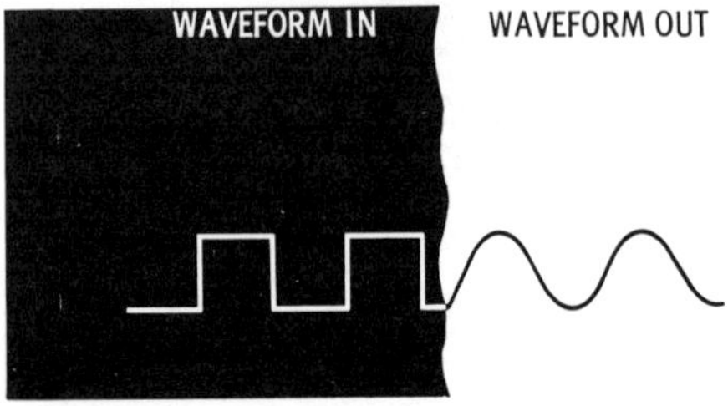


**Poor high-frequency response.**

It depends on the application of the pulse whether a particular circuit is good enough or not. In a telegraph system, it may only be necessary to detect the presence or absence of pulses with no concern as to their exact shape. So, the high-frequency response of the circuit would not be critical.

If precise location of the start of a pulse is necessary (as in a precision radar, or a timing circuit), then the high-frequency response of the circuit is extremely critical.

## TIME CONSTANTS

L/R and RC time constants indicate how quickly current or voltage builds up when a sudden increase in DC voltage (such as a square wave) is applied to a particular combination of L and R or C and R. One time constant is the time required for voltage (or current, depending on the circuit) to reach 63% of its peak value. The percentage of the peak value can be calculated for any elapsed time if the time constant of the circuit is known. The curves for the voltage increase across a capacitor, or the current increase through an inductor, are exactly the same if the time constants of the two circuits are the same.

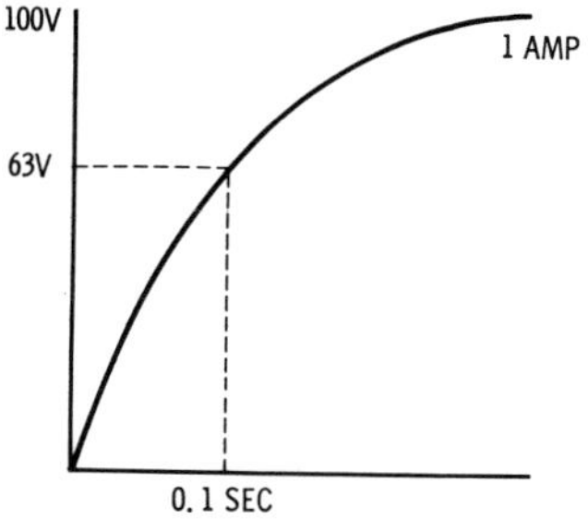


**Capacitor voltage when RC equals 0.1 second.**

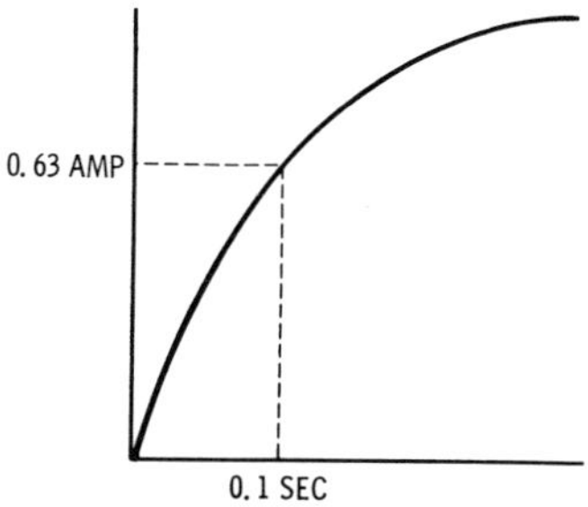


**Inductor current when L/R equals 0.1 second.**

**Q31. If a square wave is applied to the circuit below, the capacitor will tend to bypass what frequencies? What frequencies will be emphasized in the voltage across the capacitor? How will this affect the shape of the output waveform?**

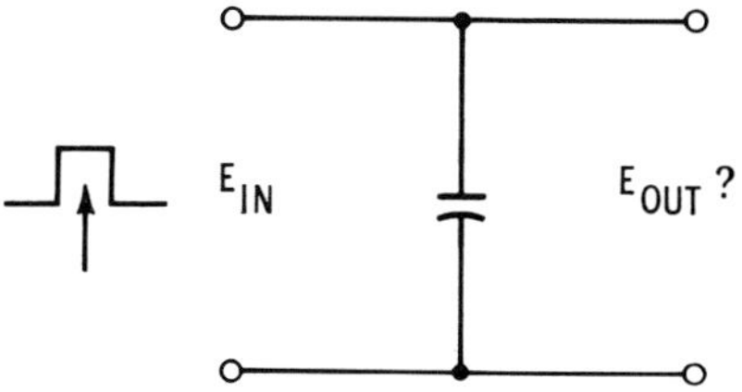


**Q32. Sketch how you think a sawtooth wave should be affected by a circuit with a poor low-frequency response. With a poor high-frequency response.**

**Your Answers Should Be:**

**A31.** The capacitor will tend to **bypass** the **high frequencies** and **emphasize** the **low frequencies** in the output voltage. This will tend to **round off the steep edges** of the waveform.

**A32.** The response to a sawtooth is very similar to that of the square wave.

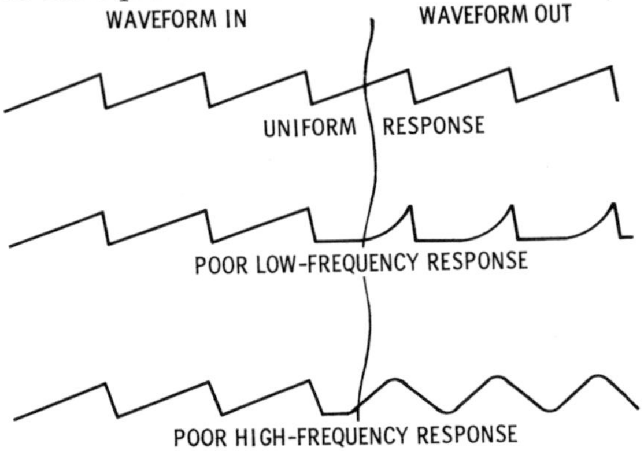


Poor low-frequency response produces a **sag** in the **sloped portions** of the wave, while poor high-frequency response **rounds off** the **sharp corners** of the wave.

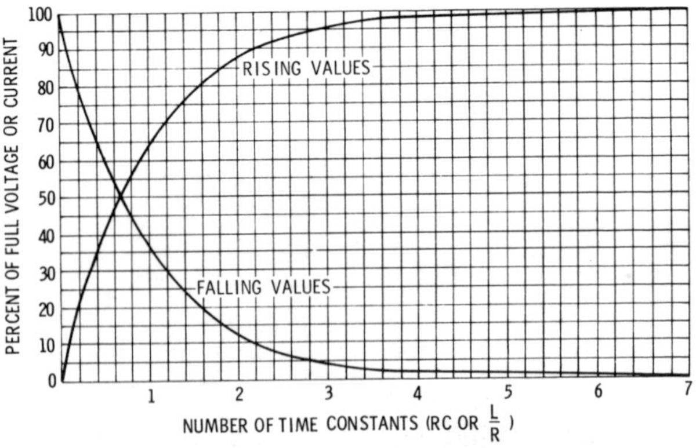


**Universal time-constant chart.**

A **universal time-constant chart** can be used to calculate the growth and decay of voltage and currents in any RL or RC circuit to which a sudden step voltage is applied. All you have to know is the final voltage and current, and the time constant of the circuit.

Calculate the voltage across the capacitor and the current in the following circuit:

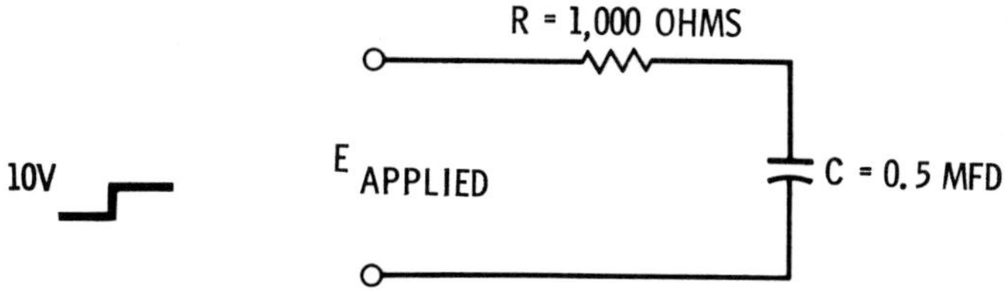


The time constant is:

$RC = 1{,}000 \times 0.5 \times 10^{-6} = 0.5 \times 10^{-3} = 0.5$ millisecond

When the 10 volts is first applied, the current will be E/R $= \frac{10}{1{,}000} = 0.01$ ampere. Using the time constant and the falling curve in the chart, it can be seen that at the end of one time constant (0.5 millisec), the current is about 37% of its full value ($0.37 \times 0.01 = 0.0037$ ampere). At the end of two time constants (1 millisecond), current will be 13% ($0.13 \times 0.01 = 1.3$ milliamperes). At the end of three time constants, current will be 5% ($0.05 \times 0.01 = 0.5$ milliampere), and so on.

The voltage across the capacitor can be determined in a similar manner. When the switch is first closed, $E_C$ is zero. This voltage gradually increases to the value of the power supply, or 10 volts. At the end of one time constant (0.5 millisec) $E_C$ will be 63% of maximum (6.3 volts). At the end of two time constants (1 millisec) $E_C$ will be 87% of maximum (8.7 volts), etc.

The same chart can be used when the voltage is suddenly removed (the switch is opened). In this case, $E_C$ decays according to the falling curve.

**Q33. Use the universal time-constant chart to describe in detail what happens when a step voltage of 10 volts is applied across a 100-ohm resistance in series with a 5-millihenry inductance.**

**Your Answers Should Be:**

**A33.**

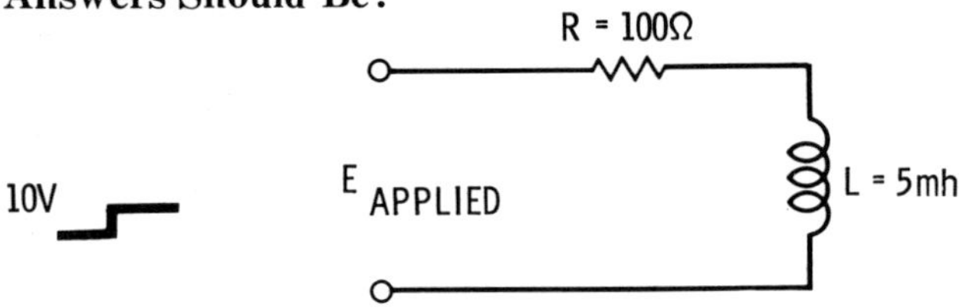


$$\text{Time constant} = \frac{L}{R} = \frac{0.005}{100} = 50 \text{ microseconds.}$$

(1) Current through R and L is zero to start with, and increases to $\frac{E}{R} = \frac{10}{100} = 0.1$ ampere eventually. Use the rising curve.

(2) Voltage across L is 10 volts initially, and drops to zero eventually. Use the falling curve.

(3) Voltage across R is zero at first, and rises to 10 volts eventually. Use the rising curve.

| Time in sec | No. of time constants | I amps | $E_L$ volts | $E_R$ volts |
|---|---|---|---|---|
| 0 | 0 | 0 | 10.0 | 0 |
| 0.25 | 0.5 | 0.038 | 6.2 | 3.8 |
| 0.35 | 0.7 | 0.050 | 5.0 | 5.0 |
| 0.50 | 1.0 | 0.063 | 3.7 | 6.3 |
| 1.00 | 2.0 | 0.087 | 1.3 | 8.7 |
| 1.50 | 3.0 | 0.095 | 0.5 | 9.5 |
| 2.00 | 4.0 | 0.098 | 0.2 | 9.8 |
| 2.50 | 5.0 | 0.099 | 0.1 | 9.9 |
| 3.00 | 6.0 | 0.0995 | 0.05 | 9.95 |
| | | 0.100 | 0 | 10.00 |

Remember that a universal time-constant chart can only be used with step voltages or square waves.

## WHAT YOU HAVE LEARNED

1. Equal amounts of capacitive and inductive reactance cancel each other when they are combined in series. If inductive reactance is greater than capacitive reactance, the total circuit will behave as though it had only an inductive reactance equal to the difference between the two reactances. If capacitive reactance is greater, the circuit will be capacitive.

2. For any series RLC circuit there is one frequency at which the two reactances are exactly equal. This is called the resonant frequency. This frequency can be found by setting the formulas for inductive and capacitive reactance equal to each other:

$$f_o = \frac{1}{2\pi\sqrt{LC}}$$

3. Maximum current will flow in a series RLC circuit at the resonant frequency, and this current will decrease at higher or lower frequencies.

4. A parallel RLC circuit has maximum impedance at its resonant frequency, and much lower impedance at higher or lower frequencies.

5. The Q of a resonant circuit is the amount of reactance of either kind, divided by the resistance. $Q = \frac{X_L}{R} = \frac{X_C}{R}$. The Q of a resonant circuit determines how quickly current or impedance decreases as the frequency is changed from the resonant frequency. High Q means a very sharp drop, low Q means a slower decrease.

6. Formulas $\frac{L}{CR} = Z_o$, $X_LQ = Z_o$, and $2\pi fLQ = Z_o$ give the impedance of a parallel resonant circuit at its resonant frequency.

7. $P = EI \cos \theta$ is a power formula that can be used to find true power in any kind of RLC circuit. Cos $\theta$ is the cosine of the phase angle between the reactive and resistive vectors, and is called the power factor. $I \cos \theta$ represents the resistive portion of the overall current.

8. The straight-line (flat-top) portions of pulses are determined by the low-frequency, sine-wave components and the steep edges are determined by the high-frequency components.
9. RLC circuits that tend to filter out low frequencies cause the straight-line portions of pulse waveforms to sag, while those that filter out high frequencies cause the steep edges to be rounded off.
10. A universal time-constant chart can be used to analyze the response of either an RL or an RC circuit to a step voltage.

# 14

# Transformer Action

## *What You Will Learn*

**Your understanding of AC electricity will now be used to show how two common electrical devices work. You will learn how a transformer transfers power from one winding to another. You will learn how to calculate the change in voltage, current, and impedance produced by a transformer with a known turns ratio, and how to select the proper turns ratio to produce a particular change. You will also learn how a magnetic amplifier controls a large AC current with a smaller DC current.**

### WHAT IS A TRANSFORMER?

A **transformer** is a device for changing the voltage of AC electricity. Transformers work on the principle of induction. Basically, a transformer has two windings—a primary and a secondary—wound on the same core. This core can be laminated iron, ferrite, or air.

Through the principle of **induction** the alternating current flowing through the primary winding sets up an alternating magnetic field in the core. This magnetic field, in turn, **induces** an alternating voltage in the secondary winding (or windings). In this way, **energy is transferred** from the primary to the secondary.

A transformer that reduces the voltage in a circuit is called a **step-down** transformer. This is true, for example, of a radio-receiver filament transformer, which steps the 120-volt main supply down to 6.3 volts.

A transformer that is used to increase the voltage in the circuit is known as a **step-up** transformer. An example is the high-voltage transformer which produces the several thousand volts needed to operate a television picture tube.

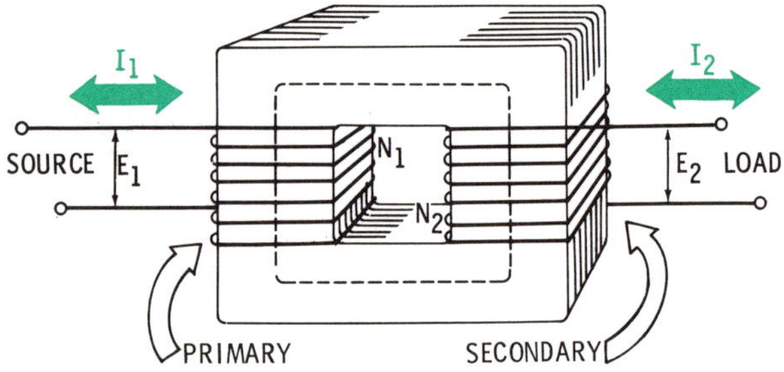


**A basic transformer.**

The basic transformer has two windings—primary and secondary—wound on a laminated-iron core. The two windings are insulated from each other and from the core.

The primary winding is connected to the energy source, and the secondary winding is connected to the load. As alternating current flows through the primary, a pulsating magnetic field is set up in the core. As the constantly changing magnetic field cuts the turns of the secondary, a voltage is induced in the secondary winding.

The amount of voltage induced in the secondary winding depends on how many turns of wire the secondary contains compared to the number of turns of wire in the primary winding. So, if the secondary has only half as many turns as the primary winding, the voltage will be stepped down to half its original value. If the secondary has twice as many turns as the primary, the voltage will be stepped up to twice its original value.

The difference in the number of turns is known as the **turns ratio** of the transformer. If the primary winding has $N_1$ turns and its voltage is $E_1$, the secondary winding with $N_2$ turns produces voltage $E_2$.

$$\frac{E_1}{E_2} = \frac{N_1}{N_2}$$

The power consumed in the secondary circuit of a transformer must be supplied by the primary. Since the voltages are constant in each circuit, the current in the primary circuit must vary to supply the amount of power demanded by the secondary. Current in the primary depends on the current drawn in the secondary circuit.

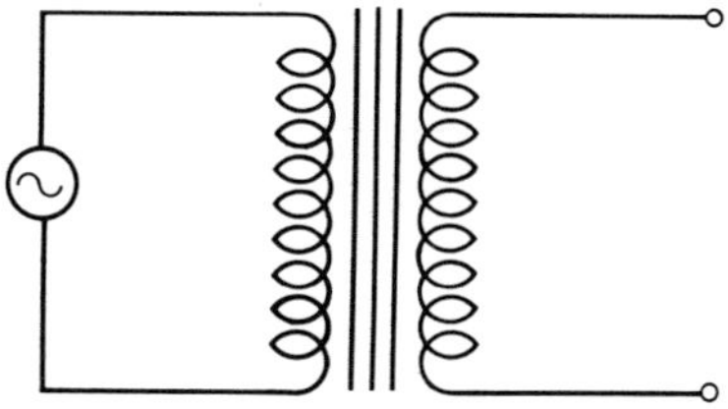

**A transformer with no load acts like an inductor.**

**Q1. If a transformer primary has 1,000 turns and the secondary has 6,500 turns, what is the turns ratio?**

**Q2. If 85 volts is applied to the primary winding of the transformer in Question 1, what is the voltage at the secondary?**

**Q3. What would happen if the leads were reversed and 85 volts was applied to the 6,500-turn coil?**

**Q4. What happens if 130 volts is fed into the 6,500-turn winding of the transformer?**

**Q5. Can a transformer be used with DC? Why?**

**Q6. What will be the phase relationship between the voltage across the primary of a transformer and the voltage across the secondary, assuming the coils are wound in the same direction?**

**Q7. If there is no load between the terminals of the secondary of a transformer, will current flow in the secondary?**

**Q8. Will there be a magnetic field produced by current in the secondary?**

**Q9. Will there be a magnetic field produced by current in the primary?**

**Q10. What effect will this magnetic field have on the impedance of the primary circuit?**

**Q11. If the magnetic field were weaker, would more or less current flow in the primary circuit?**

**Your Answers Should Be:**

**A1.**

$$\frac{N_1}{N_2} = \frac{1,000}{6,500} = \mathbf{1 \text{ to } 6.5}$$

A2.

$$\frac{E_1}{E_2} = \frac{1}{6.5}$$

$$E_2 = E_1 \times 6.5 = 85 \times 6.5 = \mathbf{552.5 \text{ volts}}$$

**A3.** If you reverse the leads, the turns ratio is:

$$\frac{N_1}{N_2} = \frac{6,500}{1,000} = \frac{E_1}{E_2}$$

**The output would be:**

$$E_2 = \frac{1,000 \times 85}{6,500} = 85 \times 0.153$$

$$= \mathbf{13 \text{ volts (approx.)}}$$

**A4.** The voltage appearing at the 1,000-turn winding will be $\frac{1,000}{6,500} \times 130 =$ **20 volts.**

**A5. A transformer cannot be used with direct current.** A direct current in the primary does not produce a pulsating magnetic field.

**A6.** The voltage across the secondary will be **180° out of phase** with the voltage across the primary.

**A7. No current will flow.**

**A8.** If no current flows, **no magnetic field** will be produced by the secondary.

**A9. Yes.**

**A10.** The **stronger the magnetic field,** the **greater will be the impedance** of the primary circuit.

**A11. More** current would flow in the primary.

## TRANSFORMER POWER

If the transformer were 100% efficient, all the power from the primary winding would be transferred to the secondary and delivered to the load.

Suppose a transformer has 1,000 turns in the primary and 6,500 turns in the secondary. If 100 volts is applied to the primary, 650 volts will appear at the secondary. Now, suppose the load connected to the secondary is a 65-ohm resistor. It will draw a current of $\frac{650}{65}$, or 10 amperes, and the power consumed will be 650 $\times$ 10, or 6,500 watts. This power must be supplied by the primary winding. Assuming no loss in the transformer, the primary winding must supply 6,500 watts. The primary current, therefore, will be $\frac{6{,}500 \text{ watts}}{100 \text{ volts}}$ $= 65$ amperes.

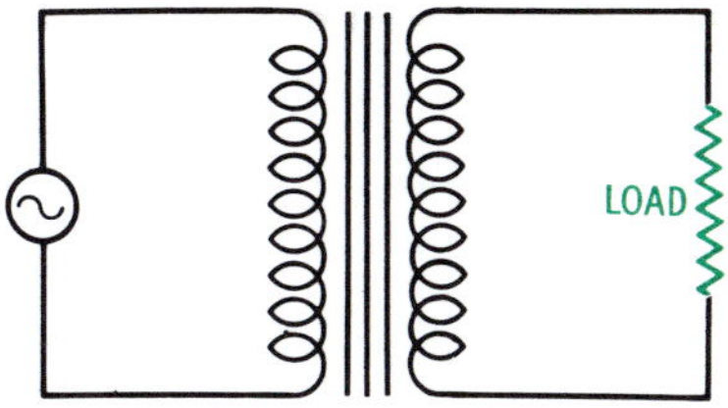


**A transformer with a load.**

In the example above, the current was stepped down in exactly the same proportion as the voltage was stepped up. The power transferred from the primary to the secondary does not change, however, regardless of the turns ratio. This is true providing the rating of the transformer has not been exceeded and assuming 100% efficiency.

**Q12. What happens to current in the secondary of a transformer when a load is connected across its terminals?**

**Q13. Will a magnetic field be produced by the current in the secondary?**

**Q14. Will the magnetic field add to or oppose the magnetic field produced by the primary? (Remember the coils are wound in the same direction but the currents are in opposite directions.)**

**Q15. How will the magnetic field produced by current flow in the secondary affect the current drawn by the primary?**

**Q16. What will happen if the load resistance in the circuit above is increased to 6,500 ohms?**

**Your Answers Should Be:**

**A12.** There **will be a current** in the secondary when a load is connected across its terminals.

**A13.** A magnetic field **will be produced** by a current in the secondary.

**A14.** The secondary magnetic field will **oppose** that of the primary.

**A15.** The secondary magnetic field will decrease the total magnetic field acting on the primary and, therefore, will decrease the impedance of the primary circuit. The **primary will draw more current.**

**A16.** Current in the secondary will be $\frac{650 \text{ volts}}{6{,}500 \text{ ohms}} =$ **0.1 ampere.** Power dissipated in the secondary will be 0.1 ampere $\times$ 650 volts = **65 watts.** Therefore, power drawn in the primary must be 65 watts. The current in the primary will then be $\frac{65 \text{ watts}}{100 \text{ volts}}$ = **0.65 ampere.**

## TRANSFORMER EFFICIENCY

So far we have assumed that no power is lost in the transfer from the primary winding to the secondary winding. However, no transformer has absolutely 100% efficiency. Some power is lost in heating the core, and some is lost in the resistance of the windings. But, transformers are very efficient; their efficiency often reaches very nearly 100%. Therefore, for rough calculations, it is permissible to assume 100% efficiency.

As with any other device, the efficiency of a transformer is equal to:

$$\frac{\text{output power}}{\text{input power}}$$

Most transformers have an efficiency in the range of 97 to 99%. So even if you neglect the losses, your calculations using 100% as the transformer efficiency will still be accurate within 1 to 3%.

## TRANSFORMER LOSSES

The power loss in transformers is due to three factors. The first is simply **resistance** in the windings; no winding is a perfect conductor. The second factor that causes power loss in transformers is **eddy currents.** The iron in the core of a transformer is a conductor. When the changing magnetic field produced by the primary coil cuts through the iron of the core, small currents are generated in the core material. These currents dissipate power as they pass through the resistance of the iron. These currents are called **eddy currents.** This type of loss is held to a minimum by using thin sheets of iron, called **laminations,** in the core. These thin sheets are insulated from each other (often by oxidizing the surface of the sheets) and thus shorten the conducting path for the eddy currents.

The third factor that causes power loss in transformers is **hysteresis.** It takes a certain small amount of power to magnetize a piece of iron. This power must be expended again when the magnetic field is reversed. Since the magnetic field in a transformer is reversed many times each second, these tiny expenditures of power add up to a noticeable loss. Hysteresis loss can be reduced by constructing the core with a type of iron that is very easily magnetized and demagnetized.

**Q17. If a transformer supplies 1.9 amperes at 100 volts to a resistive load in the secondary circuit, and if it dissipates 200 watts of power in the primary circuit, what is the efficiency of the transformer?**

**Q18. This transformer has a relatively (high, low) efficiency.**

**Q19. If the secondary of a transformer supplies 0.99 watt at 1,000 volts and the transformer has an efficiency of 99%, what power will the primary draw at 120 volts?**

**Q20. How could you find the amount of power lost due to resistance in a transformer?**

**Q21. Does an air-core transformer have hysteresis or eddy currents?**

**Your Answers Should Be:**

**A17.** The power dissipated in the secondary will be 1.9 × 100 = 190 watts. The efficiency of the transformer will be 190/200 = **95%.**

**A18.** It has a relatively **low** efficiency. (An efficiency below approximately 97% is considered to be low.)

**A19.** The voltages have no effect on the problem. The efficiency of the transformer is equal to output power divided by input power.

$$\frac{0.99}{?} = 99\%$$

The input power must be **1 watt.**

**A20.** You would have to **measure the resistance** of both windings and then calculate the **power dissipated** due to the current in the windings.

**A21.** An air-core transformer has neither **eddy-current nor hysteresis losses.**

## TYPES OF TRANSFORMERS

There are many varieties of transformers, ranging from huge power-station units to tiny subminiature radio-frequency types.

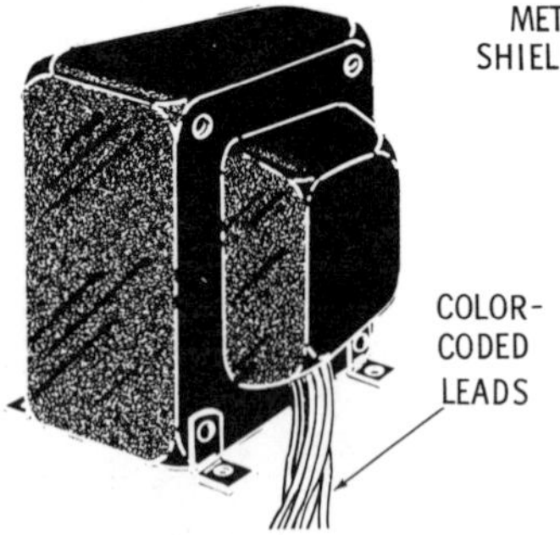


**Power-supply transformer.**

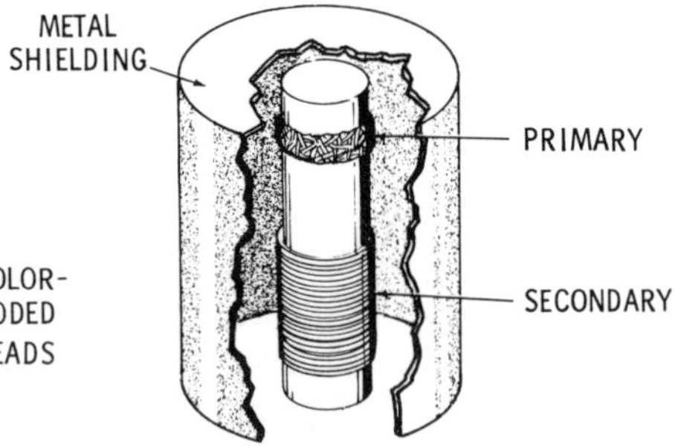


**Radio-frequency transformer.**

Most transformers are designed to transfer power. Others, however, are built to transfer only signal voltages.

Power-distribution transformers are rated in KVA (kilo-volt-amperes) rather than in kilowatts or other power units. The KVA rating refers to the apparent power carried by the transformer—the real power is smaller by the load power factor.

Special transformers, wound to precision specifications, are used in metering applications to measure the current and voltage passing through large power-transmission lines.

A step-up transformer increases voltage (which increases impedance) and decreases current (resulting from an increased impedance) at the same time. A step-down transformer decreases voltage (which decreases impedance) and increases current (which results from a decreased impedance) at the same time. Therefore, a transformer changes impedance, but the impedance change is more pronounced than the voltage change. In fact, a transformer changes impedance by the square of the turns ratio:

$$\frac{Z_1}{Z_2} = \frac{N_1^2}{N_2^2}$$

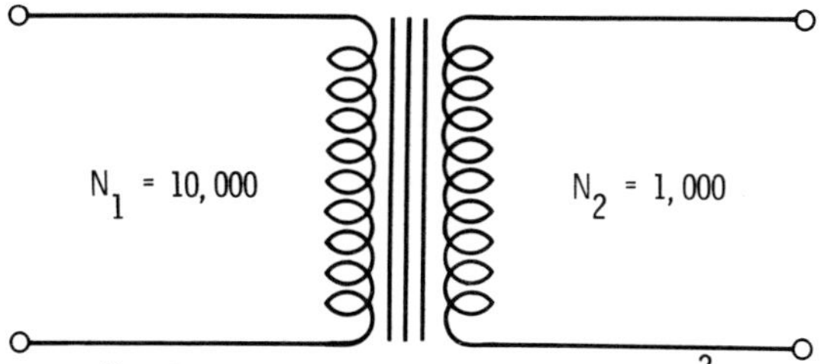


$$\frac{N_1}{N_2} = \frac{E_1}{E_2} = \frac{I_2}{I_1} \qquad \frac{Z_1}{Z_2} = \frac{N_1^2}{N_2^2}$$

**An impedance matching transformer.**

**Q22. If the primary of a transformer has 10,000 turns and the secondary has 1,000 turns, what is the turns ratio?**

**Q23. If 100 volts is applied to the primary, what voltage will appear at the secondary?**

**Q24. If the load impedance of the secondary circuit is 1 ohm, how much current will flow in the primary?**

**Q25. What is the impedance of the primary?**

**Your Answers Should Be:**

A22. $\frac{10{,}000}{1{,}000}$ = **10-to-1 turns ratio**

A23. $\frac{N_1}{N_2} = \frac{10}{1} = \frac{100}{E_2}$; $E_2 =$ **10 volts**

A24. Current in the secondary is $\frac{10}{1}$ = 10 amperes.

$$\frac{N_1}{N_2} = \frac{I_2}{I_1}; \quad \frac{10}{1} = \frac{10}{I_1}$$

$I_1 =$ **1 ampere**

A25. The impedance of the primary circuit is:

$\frac{E_1}{I_1} = \frac{100}{1}$ = **100 ohms**

## MAGNETIC AMPLIFIERS

Magnetic amplifiers are special transformer-like devices that use a small amount of power to control larger amounts of power, thus acting as amplifiers. They are simple, rugged, and efficient as compared to other forms of amplification. The following are some of the symbols used to denote magnetic amplifiers.

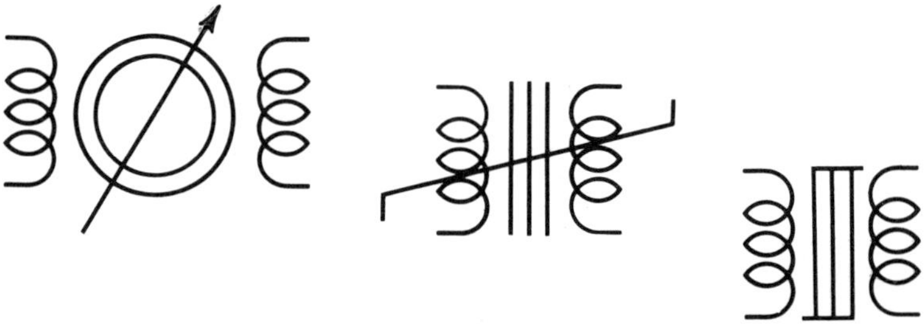

**Magnetic-amplifier symbols.**

Magnetic amplifiers take advantage of a special property of iron or steel in a strong magnetic field. To explain how a simple magnetic amplifier works, let's first review the basic principles of a coil.

When a current flows in a coil, a magnetic field (flux) is set up inside and around the coil. If the current is AC, the field also alternates. But, in any case, the strength of the magnetic field (the number of lines of flux produced) depends on the material inside the coil as well as how much current is flowing through the coil.

A very simple type of magnetic amplifier is based on the fact that an iron core normally allows greater changes in the magnetic field and, therefore, increases the inductive reactance of a coil at a given frequency.

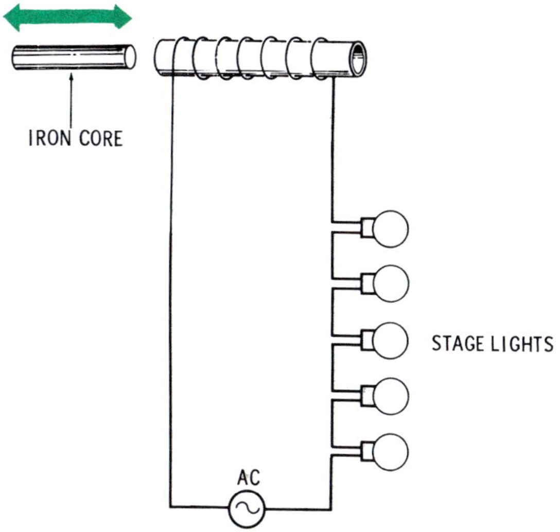


**A magnetic amplifier used to control stage lights.**

**Q26. A coil with an air core has a (greater, smaller) inductance than a similar coil with an iron core.**

**Q27. Inductive reactance is the result of a (constant, changing) magnetic field.**

**Q28. How would you increase the inductive reactance of the device illustrated above?**

**Q29. What effect would increasing $X_L$ have on the brightness of the lights?**

**Q30. How should the core be set to obtain maximum brightness of the lights?**

**Q31. Would this device work with a DC power supply?**

**Your Answers Should Be:**

**A26.** A coil with an air core has a **smaller** inductance than an iron-core coil.

**A27.** Inductive reactance is the result of a **changing** magnetic field.

**A28.** Push the iron core **into the coil.**

**A29.** Increasing $X_L$ would **dim** the lights.

**A30.** The iron core should be **totally removed.**

**A31.** The device would **not work** with DC.

## Magnetic-Amplifier Applications

A more typical magnetic amplifier controls the magnetic properties of the core and the $X_L$ of the coil by electrical means. To more easily understand how this can be done, look at the curve of current versus magnetic flux in iron.

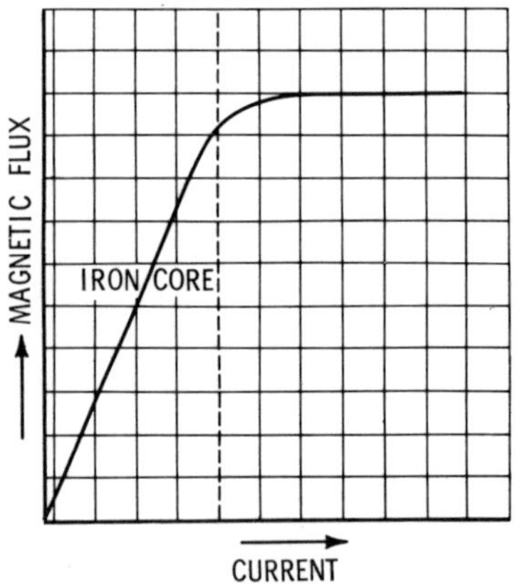


**An iron core becomes saturated as current is increased.**

As the current in the coil and the flux in an iron core increase, a point is reached where the curve bends. A further increase in the current produces less and less of a flux increase until, finally, a further increase in current produces no additional flux. At this point the core contains as many lines of flux as it can possibly receive; it is said to be **saturated.**

To make a magnetic amplifier, wind two different coils on a core in the same manner as a transformer is constructed.

With no current in the **control winding,** the current flowing through the **load winding** is limited by a strong reactive impedance. The light will shine very dimly, if at all. But if enough DC current is passed through the control winding to saturate the core, then as the saturation point is approached, the impedance of the load winding decreases, and the light will become brighter.

Since the control winding uses less power than the load winding, we have an amplifier—a device in which a small amount of power controls a large amount of power.

As AC current flows through the load winding, an AC voltage is induced back into the control winding. This is a loss of power. The control winding is made immune to the induced voltage by using a **three-legged core.**

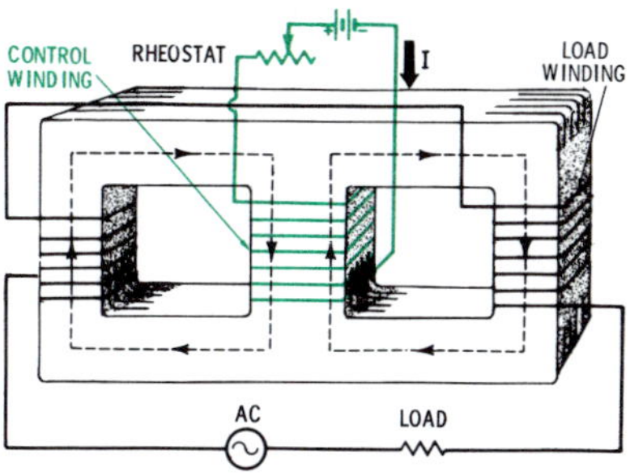


**AC current flow through the load is controlled by DC current in a control winding.**

The two parts of the load winding are connected in series in such a way that the AC flux lines they induce in the center leg are equal and opposite, and thus cancel each other. This means there is no AC induced in the control winding, yet the control winding still exerts its influence over the load circuit.

**Q32. A coil with an unsaturated iron core has a relatively (large, small) $X_L$.**

**Q33. If magnetic flux is added from an outside source (such as a separate coil on the same core) until the core is saturated, will the coil have a higher or lower $X_L$?**

**Q34. What type of current is affected by the coil $X_L$?**

**Your Answers Should Be:**

A32. A coil with an unsaturated iron core has a relatively **large** $X_L$.

A33. If the core is saturated, $X_L$ will be **lower.**

A34. **Only AC** is affected by inductive reactance.

## WHAT YOU HAVE LEARNED

1. The changing magnetic field produced by the primary winding in a transformer induces a changing voltage in the secondary winding.
2. The ratio of the primary voltage to the secondary voltage is the same as the ratio of the number of turns in the primary winding to the number of turns in the secondary winding.
3. If a transformer steps up voltage, it steps down current, and vice versa. The power drawn by the primary winding is equal to the power dissipated in the secondary circuit.
4. Most transformers have an efficiency of nearly 100%, so very little power is lost in them.
5. Transformers alter the impedance of a load. The change in impedance depends on the square of the turns ratio.
6. Magnetic amplifiers control the inductive reactance of a coil by altering the magnetic property of its core.
7. A very simple magnetic amplifier is basically a coil with a removable iron core. When the core is inserted, $X_L$ increases, and the power supplied to the load decreases.
8. Iron cores can receive only a certain amount of magnetic flux, and when they are saturated, they no longer increase the $X_L$ of a coil.
9. A typical magnetic amplifier uses a DC current through a control winding to control the level of saturation of an iron core and thus the $X_L$ of an AC load winding.

# NOTES

## NOTES

# NOTES

# NOTES